T0094129

Computational Trust Models and Machine Learning

Chapman & Hall/CRC
Machine Learning & Pattern Recognition Series

SERIES EDITORS

Ralf Herbrich
Amazon Development Center
Berlin, Germany

Thore Graepel
Microsoft Research Ltd.
Cambridge, UK

AIMS AND SCOPE

This series reflects the latest advances and applications in machine learning and pattern recognition through the publication of a broad range of reference works, textbooks, and handbooks. The inclusion of concrete examples, applications, and methods is highly encouraged. The scope of the series includes, but is not limited to, titles in the areas of machine learning, pattern recognition, computational intelligence, robotics, computational/statistical learning theory, natural language processing, computer vision, game AI, game theory, neural networks, computational neuroscience, and other relevant topics, such as machine learning applied to bioinformatics or cognitive science, which might be proposed by potential contributors.

PUBLISHED TITLES

BAYESIAN PROGRAMMING
Pierre Bessière, Emmanuel Mazer, Juan-Manuel Ahuactzin, and Kamel Mekhnacha

UTILITY-BASED LEARNING FROM DATA
Craig Friedman and Sven Sandow

HANDBOOK OF NATURAL LANGUAGE PROCESSING, SECOND EDITION
Nitin Indurkhya and Fred J. Damerau

COST-SENSITIVE MACHINE LEARNING
Balaji Krishnapuram, Shipeng Yu, and Bharat Rao

COMPUTATIONAL TRUST MODELS AND MACHINE LEARNING
Xin Liu, Anwitaman Datta, and Ee-Peng Lim

MULTILINEAR SUBSPACE LEARNING: DIMENSIONALITY REDUCTION OF MULTIDIMENSIONAL DATA
Haiping Lu, Konstantinos N. Plataniotis, and Anastasios N. Venetsanopoulos

MACHINE LEARNING: An Algorithmic Perspective
Stephen Marsland

A FIRST COURSE IN MACHINE LEARNING
Simon Rogers and Mark Girolami

MULTI-LABEL DIMENSIONALITY REDUCTION
Liang Sun, Shuiwang Ji, and Jieping Ye

ENSEMBLE METHODS: FOUNDATIONS AND ALGORITHMS
Zhi-Hua Zhou

Chapman & Hall/CRC
Machine Learning & Pattern Recognition Series

Computational Trust Models and Machine Learning

Edited by

Xin Liu
EPFL
Lausanne, Switzerland

Anwitaman Datta
Nanyang Technological University
Singapore

Ee-Peng Lim
Singapore Management University

CRC Press
Taylor & Francis Group
Boca Raton London New York

CRC Press is an imprint of the
Taylor & Francis Group, an **informa** business

A CHAPMAN & HALL BOOK

CRC Press
Taylor & Francis Group
6000 Broken Sound Parkway NW, Suite 300
Boca Raton, FL 33487-2742

© 2015 by Taylor & Francis Group, LLC
CRC Press is an imprint of Taylor & Francis Group, an Informa business

No claim to original U.S. Government works

Printed on acid-free paper
Version Date: 20140728

International Standard Book Number-13: 978-1-4822-2666-9 (Hardback)

Library of Congress Cataloging-in-Publication Data

Computational trust models and machine learning / editors, Xin Liu, Anwitaman Datta, Ee-Peng Lim.
    pages cm -- (Chapman & Hall/CRC machine learning & pattern recognition series)
    Includes bibliographical references and index.
    ISBN 978-1-4822-2666-9 (hardback)
    1. Computational intelligence. 2. Machine learning. 3. Truthfulness and falsehood--Mathematical models. I. Liu, Xin (Mathematician) II. Datta, Anwitaman. III. Lim, Ee-Peng.

Q342.C675 2014
006.3'1--dc23
                                        2014028238

Visit the Taylor & Francis Web site at
http://www.taylorandfrancis.com

and the CRC Press Web site at
http://www.crcpress.com

To my parents and my wife
— Xin Liu

Contents

List of Figures

List of Tables

Preface

Trust has implicitly played a vital role in human societies for eons. With the advent of the digital age, and the bulk of our day-to-day activities and interactions shifting online, there is a natural need to devise mechanisms to infer trust in the brave new cyberworld. Unlike a physical interaction, online interactions have many distinctive aspects, some of these hinder the decision-making process, while others may be harnessed to improve it. For instance, for e-commerce, our physical senses to inspect the goods are not applicable, nor are mechanisms to redress problems in a face-to-face interaction feasible, since the interactions may happen among geographically dispersed people. At the same time, digital footprints (say, of people's activities) may provide access to insight that has hitherto been unavailable in the physical world.

Trust is also inherently multifaceted. One may be trustworthy for certain aspects, and yet not be trustworthy in other aspects. This brings about the need to be able to discern the aspect on which trust is being determined, or the ability to translate or project trust from one aspect into another.

There is also an interesting duality between the trust one can build, guided by credible information, and ascertaining the credibility and trustworthiness of information one may find in an open system.

Trust has long been a subject of qualitative study among sociologists, and the insights from social science are relevant even in the digital domain. However, such studies fall short in several ways. Foremost, often, such a study is not directly useful in decision making. Secondly, the traditional approach for determining the necessary evidence to carry out the study and draw conclusions is both ineffective and arguably unnecessary. A plethora of data is readily captured for all sorts of electronic activities. There are arguably many challenges and few concrete answers yet, and this is natural — we are at the infancy of the digital age.

Nevertheless, the digital age has forced us to think of trust in a quantitative manner and to do so by pursuing a data-driven methodology. That is the underpinning of computational trust. And even though the techniques are arguably in their nascence, we have come a long way from the point where trust could be talked about only qualitatively. Machine learning and data mining techniques are natural tools which are shaping the study of computational trust, and in this book, we have tried to capture a representative set of such studies from several groups worldwide. Each chapter is an independent contribution from distinct research groups. We have aimed to cover several ideas, but this book is definitely not an exhaustive survey. This treatment of the topic is thus more suitable as companion reading material for established researchers and novice graduate students getting introduced to the field, who wish to get a quick flavor of several disparate ideas under a broad umbrella. It should not be used as a reference textbook which has synthesized all the ideas and structured them systematically. Accordingly, barring the introductory chapter, the material can be read more or less in any order, since each chapter is reasonably self-contained.

Xin Liu from École Polytechnique Fédérale de Lausanne, Switzerland, has provided an introductory survey of the traditional treatment of computational trust which does not apply machine learning techniques. This chapter also provides a critical summary of some of the drawbacks with those approaches, thus paving the way for the use of more sophisticated but robust machine learning techniques. The chapter also introduces some very basic ideas from machine learning.

In Chapter 2, Athirai Aravazhi Irissappane and Jie Zhang, both from Nanyang Technological University, Singapore, provide a broad overview of how reputation-based systems are used to determine trust in diverse kinds of online communities, and how machine learning techniques are employed to build robust reputation systems.

Chapter 3 is contributed by Jeff Pasternack from Facebook, Inc., and Dan Roth from the University of Illinois at Urbana-Champaign. Chapter 4 is contributed by researchers from the Polish-Japanese Institute of Information Technology, namely, Adam Wierzbicki, Andrzej Kostański, Bartłomiej Balcerzak, Dominik Deja, Grzegorz Kowalik, Katarzyna Gniazdik, Maria Rafalak, Marta Juźwin, Michał Kąkol and Wiesław Kopeć, in collaboration with Xin Liu from École Polytechnique Fédérale de Lausanne, Switzerland. Both of these chapters explore ways to determine the credibility of resources, typically articles, but do so following two distinctive approaches. The former studies an automated, iterative learning process where human role is implicit, while the latter leverages human input explicitly by embracing crowd-sourcing to determine content credibility.

Chapter 5 by Mohammad Ali Abbasi, Jiliang Tang and Huan Liu from Arizona State University demonstrates how decision support can be facilitated by computational trust models. To that end, they elaborate collaborative filtering-based trust-aware recommendation systems.

Arguably, all the trust models essentially incorporate explicit and/or implicit human feedback to carry out and quantify the resulting trust value(s). However, human inputs are an artifact of their personal biases. Thus, there is a need to filter out outlying opinions, even while making sure critical red flags are not ignored. We conclude the book with Chapter 6, where Hady Wirawan Lauw from Singapore Management University brings us back full circle and investigates the objectivity of this feedback itself, so that the trust model manages to leverage credible feedback, and yet filter out the noise therein.

This rich coverage made compiling this book an enjoyable enterprise for the editors. We hope it also makes a compelling read for the audience.

About the Editors

Xin Liu is a postdoctoral researcher at the Distributed Information Systems Laboratory (LSIR) of École Polytechnique Fédérale de Lausanne (EPFL). He earned his BSc from the College of Computer Science and Technology, Jilin University, China, and his PhD from the School of Computer Engineering, Nanyang Technological University (NTU), Singapore. In April 2012, he joined EPFL and works on the project *RecONCILE: Robust Online Credibility Evaluation of Web Content*. He also collaborates on other European projects such as PlanetData and Wattalyst. His research interests include recommender systems, trust and reputation and social computing. His papers have been accepted at high quality conferences and journals including the International World Wide Web Conference (WWW), AAAI Conference on Artificial Intelligence (AAAI), International Joint Conference on Artificial Intelligence (IJCAI), International Conference on Information and Knowledge Management (CIKM), International Conference on Autonomous Agents and Multiagent Systems (AAMAS), Electronic Commerce Research and Applications (ECRA), Computational Intelligence, etc.

School for Computer and Communication Sciences
École Polytechnique Fédérale de Lausanne
Lausanne, Switzerland
Email: x.liu@epfl.ch

Anwitaman Datta joined NTU Singapore in 2006 after earning his PhD from EPFL Switzerland. He is interested in large-scale networked distributed information systems and social collaboration networks, particularly the reliability and security of such complex systems. He leads the Self-* and Algorithmic aspects of Networked Distributed Systems (SANDS) research group at NTU.

School of Computer Engineering
Nanyang Technological University
Singapore
Email: anwitaman@ntu.edu.sg

Ee-Peng Lim is a professor at the School of Information Systems of Singapore Management University (SMU). He earned PhD from the University of Minnesota, Minneapolis and BSc from the National University of Singapore. His research interests include social networks and web mining, information integration, and digital libraries. He is the director of the Living Analytics Research Center and is the Lee Kuan Yew Fellow for Research Excellence at SMU. He is currently the associate editor of several international journals including the *ACM Transactions on Information Systems* (TOIS), *ACM Transactions on the Web* (TWeb), *IEEE Transactions on Knowledge and Data Engineering* (TKDE), *Information Processing and Management* (IPM), *Social Network Analysis and Mining*, and *IEEE Intelligent Systems*. He serves on the steering committee of the International Conference on Asian Digital Libraries (ICADL), Pacific Asia Conference on Knowledge Discovery and Data Mining (PAKDD), and International Conference on Social Informatics (Socinfo).

School of Information Systems
Singapore Management University
Singapore
Email: eplim@smu.edu.sg

Contributors

Mohammad Ali Abbasi
Arizona State University
Tempe, Arizona

Bartłomiej Balcerzak
Polish-Japanese Institute of
    Information Technology
Warsaw, Poland

Dominik Deja
Polish-Japanese Institute of
    Information Technology
Warsaw, Poland

Katarzyna Gniadzik
Polish-Japanese Institute of
    Information Technology
Warsaw, Poland

Athirai A. Irissappane
Nanyang Technological University
Singapore

Marta Juźwin
Polish-Japanese Institute of
    Information Technology
Warsaw, Poland

Michał Kąkol
Polish-Japanese Institute of
    Information Technology
Warsaw, Poland

Wiesław Kopeć
Polish-Japanese Institute of
    Information Technology
Warsaw, Poland

Andrzej Kostański
Polish-Japanese Institute of
    Information Technology
Warsaw, Poland

Grzegorz Kowalik
Polish-Japanese Institute of
    Information Technology
Warsaw, Poland

Hady Wirawan Lauw
Singapore Management University
Singapore

Xin Liu
École Polytechnique Fédérale de
    Lausanne
Lausanne, Switzerland

Huan Liu
Arizona State University
Tempe, Arizona

Jeff Pasternack
Facebook, Inc.
Menlo Park, California

Maria Rafalak
Polish-Japanese Institute of
    Information Technology
Warsaw, Poland

Dan Roth
University of Illinois at
    Urbana-Champaign
Urbana, Illinois

Jiliang Tang
Arizona State University
Tempe, Arizona

Adam Wierzbicki
Polish-Japanese Institute of
   Information Technology

Warsaw, Poland

Jie Zhang
Nanyang Technological University
Singapore

Chapter 1

Introduction

1.1 Overview

The advent of the Internet and the Web enables people to strongly connect with each other. Due to the openness of various online applications, end-users play active role in the system, and thus their activities in turn significantly impact the system's behavior and other users' experiences. For instance, different from traditional static Websites where information is just displayed to users, Web 2.0 applications allow users to not only extract information but also contribute and share user generated contents with others (e.g., Wikipedia, Twitter, Slideshare, Facebook).

The intensive interactions between users in online applications, on the one hand, significantly enhance user experience, but on the other hand, also create the issues of security and reliability. Dishonest users can easily join the system and behave maliciously or selfishly to achieve their goals while compromising honest users' experiences. Such security implications raise an important research question: *How can we ensure that the (unknown) potential interaction partners will not harm the interests of normal users in open systems?*

Traditional security solutions such as Authentication, Authorization and Accounting (AAA) [128] and Public Key Infrastructure (PKI) [284] indeed help to reduce the impact of behavior of the dishonest users. However, these strategies only help to make sure that the interaction partner is authenticated and authorized to take actions, but they do not guarantee that the partner is doing what is expected, i.e., they only assume complete certainty thus belonging to "hard security". In this scenario, a "soft" approach is needed to deal with the uncertainty of users' behavior (e.g., we never know other users' intentions; we cannot tell if the user we are going to interact with has sufficient competence and resources) that is quite common in massively distributed systems. Trust, in this case, is a kind of "soft security" we need.

The rest of this chapter is organized as follows: In Section 1.2, we formally discuss the concept of trust and other related concepts such as reputation. In Section 1.3, we elaborate on computational trust modeling where we first review traditional trust models in Section 1.3.1 and then in Section 1.3.2, emphasize the recently emerged supervised learning based trust modeling. Fundamentals of machine learning are introduced in Section 1.3.2.1 and novel supervised learning based approaches are discussed in Section 1.3.2.2. Finally, we provide the organization of this book in Section 1.4.

1.2　What Is Trust?

The concept of trust is prevalent in our society. From medical treatment to business trade, we must establish a certain level of trust before acting. Particularly, with the development of the Internet and World Wide Web (WWW), people increasingly interact with others in the virtual world, and the need for trust is thus widespread in various areas of the digital world. For instance, in online auction sites, before performing a transaction, the buyer must ensure that the seller will not cheat on the transaction such as selling low quality item.

Typically, trust is referred to as the relationship between two entities, where one entity (trustor) is willing to rely on the (expected) actions performed by another entity (trustee). Sociologically, trust is attributed to relationship (1) between individual persons, (2) between a person and an object or action, (3) within and between social groups (family, companies, countries, etc.). In the domain of computer science, human trust is understood to refer to the relationships among various virtual entities (e.g., a piece of software, devices, end users, a subset of the network). We next summarize definitions of trust in literature to help readers understand trust.

According to Gambetta's definition [76] (Page 217), "Trust (or, symmetrically, distrust) is a particular level of the subjective probability with which an agent assesses that another agent or group of agents will perform a particular action, both before it can monitor such action (or independently of his capacity ever to be able to monitor or enforce it) and in a context in which it affects his own action." From this definition we can see that trust is subjective, because it is estimated from the perspective of the individual trustor.

Furthermore, trust exists before the action is observed. That is to say, trust is a prior belief about an agent's behavior. Another point that needs to be noted is that trust is in a certain context, which affects the trustor's own actions.

Marsh [164] was one of the first to define trust from a computational point of view. He agrees that trust implies some degree of uncertainty and hopefulness or optimism regarding an outcome, being subjective and dependent on the views of the individual [84]. The author categorized trust in three types: (1) Basic trust models the general trusting disposition independently of whoever is the encountered agent. This trust is actually the accumulation of direct experience. Positive direct experience leads to a greater disposition to trust, and vice versa. (2) General trust is the trust that one agent has in another agent without considering any specific situation. (3) Situational trust is the trust that one agent has in another taking into account a specific situation.

Jøsong et al. [117] identified two definitions of trust (Page 3,4): (1) Reliability trust is defined according with Gambetta [76]: "Trust is the subjective probability by which an individual, A, expects that another individual, B, performs a given action on which its welfare depends." (2) Decision trust is inspired by McKnight and Chervany [174]: "Trust is the extent to which one party is willing to depend on something or somebody in a given situation with a feeling of relative security, even though negative consequences are possible."

The definitions introduced above demonstrate various aspects of the concept of trust and are very necessary to comprehensively understand "what is trust?" It is worth noting that in the domain of trust research, reputation is a related but distinct concept. Basically, the concept of reputation is derived from people's common sense and economics [255] and is treated as one of the popular ways to build trust. According to Abdul-Rahman and Hailes [7] (Page 3), "A reputation is an expectation about an agent's behavior based on information about or observations of its past behavior." From the definition we observe that reputation can be derived from either trustor's past experience with target agent or opinions reported by third parties. In Jøsong et al. [117] (Page 5), reputation is defined as "what is generally said or believed about a person's or thing's character or standing".

Reputation connects closely to the concept of trust, but there is a clear difference, which can be illustrated by the following two scenarios:

(1) *A* trusts *B* because *B* has a good reputation. This reflects that reputation can be used to build trust.

(2) *A* trusts *B* despite *B*'s bad reputation. This reflects that even if *A* knows *B*'s reputation, *A* has its own private knowledge (e.g., direct experience about *B*), which is more important for *A* to judge *B*'s trustworthiness than reputation.

From these two scenarios, we notice that trust is subjective and personalized. Trust can be derived based on various factors, in which some of them (e.g., direct experience) are more important than others (e.g., reputation).

1.3   Computational Trust

Computational trust applies the human notion of trust to the digital world. Specifically, computational trust is a term that describes representations of trust used for trust inference (e.g., in trust management systems), and these representations can be based on diverse trust definitions. In the past 15 years, computational trust has been thoroughly studied in various computer science fields (e.g., multi-agent systems, social networks, distributed and Peer-to-Peer networks), and a large volume of computational trust models have been proposed for different application scenarios, focusing on different aspects of trust modeling. In this section, we first provide a brief review of traditional computational trust models in Section 1.3.1. Then we discuss recently emerged machine learning based trust modeling in Section 1.3.2.

1.3.1   Computational Trust Modeling: A Review

Existing online applications provide a large-scale, open and heterogeneous interaction environment where diverse types of information can be collected from different sources. One popular method to estimate trust is to rely on past behavior of the specific agent in question. The basic idea of such a method is to let agents involved in the interactions assess (e.g., rate) each other. For instance, after an interaction, the service requestor rates the service provider according to his experience in the interaction. Such ratings are used by this service requestor to determine whether or not to interact with that service provider in the future encounters. Generally, if an agent mostly performs honestly in the past, it is considered to be trustworthy in the future interactions, otherwise, it is considered to be untrustworthy.

Intuitively, direct experience is the most reliable and personalized information for trust assessment. However, in large-scale, open systems such as online social networks, direct experience is often not sufficient or even nonexistent. In this case, another way to judge the target agent's past behavior is to rely on indirect experience, which is the opinions obtained from other agents who have interacted with the target agent.

Although it may not be as accurate/reliable as direct experience, indirect experience is much more pervasive, thus greatly complementing direct experience based trust approaches [8,115]. However, in an open, complex system, it is nontrivial to understand and aggregate such information due to its uncertainty (e.g., the information reporter may provide false information, or although the provided information is correct, it may not be suitable for the information requester due to its personalized view of the system). In order to address this issue, a lot of approaches have been proposed [23, 133, 264, 288, 298], and a comparison study of social network based and probabilistic estimation based approaches was conducted in Despotovic and Aberer [54]. Some other indirect experience based trust and reputation models rely on game theoretical approaches [236,293] to predict the behavior of the interaction (game) partner. However, in a large-scale social network system, the reliability of game theoretical models decreases due to high complexity of relations and interactions among agents.

Direct experience and indirect experience are commonly used to derive trust. However, when such information is not available, one may rely on other kinds of information by simulating human perceptions such as stereotype. While using stereotypes for user modeling and decision making was suggested previously [171], StereoTrust [150] is probably the first concrete, formal computational trust model that uses stereotypes. In this work, the trustor forms stereotypes by aggregating information from the context of the transaction, or the profiles of their interaction partners. Example stereotypes are "programmers working for Google are more skilled than the average" or "people living in good neighborhoods are richer than the others". In order to build stereotypes, the trustor has to group relevant agents ("programmers working for Google" or "people living in good neighborhoods"). Stereotypes on each group are calculated by aggregating the trustor's past experience with members of that group.

When facing a new agent, the trustor uses its stereotypes on the groups to which the new agent belongs to, and the weight for each stereotype is fraction of the trustor's transactions with members of the corresponding group. Example environments where StereoTrust may be applied include (1) identifying unknown malicious sellers in online auction sites [151], (2) selecting reliable peers to store replicated data in Peer-to-Peer storage systems [154], to name a few.

After StereoTrust, several other stereotyping based works are proposed in the area of multi-agent system. Burnett et al. [28] used stereotypes to address the cold-start issue. The trustor constructs stereotypes relying on machine learning algorithm (i.e., M5 model tree learning [214]), and combine the stereotypes with the target agent's reputation (if any) using subjective logic [114]. Furthermore, the authors extended the model by allowing new trustors to request experienced trustors who have already formed local stereotypes. Similar approaches that also employ supervised learning algorithms include MetaTrust, etc. [152,153]. We will elaborate such supervised machine learning based trust models in Section 1.3.2.

Burnett et al. [29] looked at the different aspect of the problem, that is to identify useful features to construct stereotypes. Three feature sources are discussed. (i) From social network. That is, relationships between agents can be viewed as features. For instance, such relationships could be *agent A is a friend of agent B*. (ii) From agents' competence over time. The target agent's accumulated experience in certain tasks can be viewed as features. An example stereotype may be *if an agent performed task T for over 100 times, he is considered experienced (trustworthy)*. (iii) From interactions, e.g., features of both interaction parties. For instance, the trustor with certain features is positively or negatively biased towards the target agent with other features. This work provided a comprehensive summary of feature sources (for stereotype formation) from social relationships among agents but the authors did not apply these features to any concrete application scenarios for validation.

In the next subsections, we summarize some representative trust computation methods.

1.3.1.1 Summation and Average

The simplest approach to compute trust/reputation is to simply aggregate all the ratings. There are several types of aggregations. For instance, in eBay's reputation system (http://www.ebay.com), for recent 12 months, positive, neutral and negative feedback are summed separately (Figure 1.1).

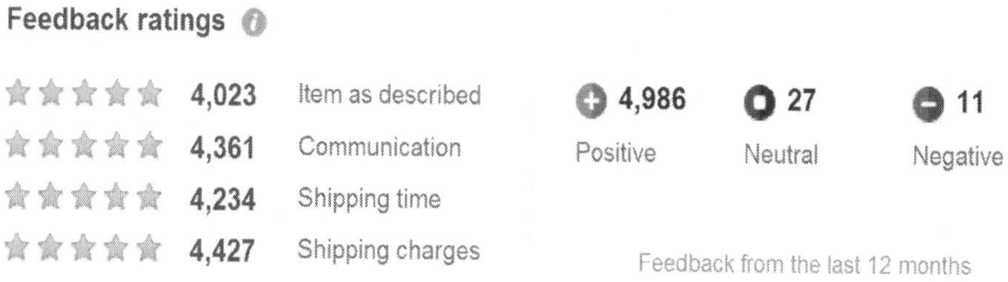

Feedback ratings ⓘ

★★★★★ 4,023	Item as described	⊕ **4,986** Positive
★★★★★ 4,361	Communication	◑ **27** Neutral
★★★★★ 4,234	Shipping time	⊖ **11** Negative
★★★★★ 4,427	Shipping charges	Feedback from the last 12 months

FIGURE 1.1: Snapshot of a seller profile on eBay.

Another approach is to compute reputation as the average of all ratings as Amazon does. Figure 1.2 is an example where the aggregation of ratings (five-point scale) is showed on each item's profile page.

7,382 Reviews

5 star: ▓▓▓ (4,958)
4 star: ▓ (1,304)
3 star: | (451)
2 star: | (274)
1 star: | (395)

Average Customer Review

★★★★☆ (7,382 customer reviews)

Share your thoughts with other customers

Create your own review

FIGURE 1.2: Snapshot of an item profile on Amazon.

1.3.1.2 Bayesian Inference

The beta reputation system proposed by Jøsang et al. [115] estimates the reputation of an agent using a probabilistic model, i.e., beta probability density function (PDF).[1] The beta distributions are a family of statistical distribution functions that are characterized by two parameters α and β. The beta probability density function is defined as follows:

$$beta(p|\alpha, \beta) = \frac{\Gamma(\alpha + \beta)}{\Gamma(\alpha)\Gamma(\beta)} p^{\alpha-1}(1-p)^{\beta-1}. \tag{1.1}$$

Here $p \in [0, 1]$ is a probability variable, and $\alpha, \beta > 0$. This function shows the relative likelihood of the values for p, given the parameters α and β. The probability expectation of the beta distribution is computed by $E(p) = \alpha/(\alpha + \beta)$.

In the beginning, when no knowledge is available, the prior distribution is actually the uniform beta PDF with the parameters $\alpha = 1$ and $\beta = 1$ (see Figure 1.3(a)). After observing x positive and y negative outcomes, the posterior distribution becomes beta PDF with $\alpha = 1 + x$ and $\beta = 1 + y$. Figure 1.3(b) gives the example when $x = 10$ and $y = 2$.

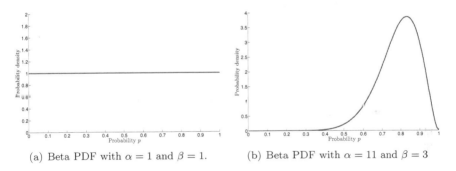

(a) Beta PDF with $\alpha = 1$ and $\beta = 1$.   (b) Beta PDF with $\alpha = 11$ and $\beta = 3$

FIGURE 1.3: Examples of beta PDF.

Following this model, we assume the feedback collected from other agents who have interacted with the target agent is binary (1 or 0), i.e., 1 means that the target agent has a good reputation and 0 otherwise. These third party opinions are combined by simply accumulating (a) the number of positive feedback and (b) the number of negative feedback. Hence, To make $\alpha, \beta > 0$, their values are set as $\alpha = 1 + a$ and $\beta = 1 + b$, respectively. The reputation score can be obtained based on the probability expectation value. The posterior reputation score is updated by combining a prior reputation score with the new feedback.

---

[1]A natural extension is to apply Dirichlet distribution, which is a generalization of beta distribution to model multiple discrete trust rating levels [113]

The beta reputation system computes and updates the target agent's reputation following beta distribution. However, since there are no effective mechanisms provided to filter out unfair responses, this work is only effective when most of the feedback is genuine. TRAVOS [264] is a trust and reputation model for agent-based virtual organizations. This work also makes use of beta distribution to compute trust but it pays more attention to the issue of unfair feedback. When the trustor evaluates a potential interaction partner, he first uses his own local knowledge to derive trustworthiness of the target agent. He then estimates accuracy of the direct trust, which is the probability that the real likelihood of cheating falls within a certain range from the trustor's estimation. If the direct trust is not accurate enough, the trustor requests feedback from other agents who have interacted with the target agent. To ensure that only accurate feedback is considered, TRAVOS addresses inaccurate reputation feedback by performing two tasks: (1) estimating the probability that a feedback provider's opinion of the target agent is accurate by comparing current feedback with the previous feedback provided by the same agent and (2) adjusting reputation feedback according to its accuracy to reduce the effect of the inaccurate feedback.

Compared to the beta reputation system [115], TRAVOS (as well as other beta distribution-based enhancements [23, 285]) takes into account the issue of unfair feedback, thus deriving more accurate reputation score of the target agent. However, TRAVOS assumes that an agent acts consistently, which is clearly not true in reality. For instance, in Internet auction sites, malicious sellers may first act honestly in some transactions selling cheap items and then cheat in a transaction selling expensive items. Moreover, both beta reputation systems and TRAVOS rely on an agent's past behavior (direct or indirect experience), which is not always available. For instance, an agent may be new to the system or an agent may interact with others infrequently.

Zhang and Cohen [298] took into account an agent's dynamic behavior by introducing the concept of time window. That is, the ratings of the target agent are partitioned into different elemental time windows. In each time window, the trustor counts the numbers of successful and unsuccessful transactions. Trustworthiness of the target agent is firstly calculated by aggregating numbers of successful and unsuccessful transactions in each time window (taking into account forgetting rate) and then is adjusted according to reputations of the indirect experience providers.

1.3.1.3   Web of Trust

Web of trust models or transitive trust models, e.g., [6], build chains of trust relationships between the trustor and the target agent. The basic idea is that if A knows B and B knows C, then A can derive C's trust using B's referral of C and A's trust in B, subject to any context constraints. A real world implementation of such a trust model can be found at Epinions[2] where

---

[2]www.epinions.com

a user's trust relationship information is displayed on her profile page (e.g., the users she trusts and those who trust her).

Abdul-Rahman and Hailes [6] proposed a distributed trust model based on conditional transitivity of trust. Consider the scenario that A trusts B and B trusts C, one can simply derive that A trusts C if the following conditions are true:

- B recommends its trust in C to A explicitly;

- A trusts B as a recommender; and

- A can judge B's recommendation and decide how it will trust C, irrespective of B's trust in C.

Abdul-Rahman and Hailes [6] summarized properties of such a transitive trust relationship: (1) it is always between exactly two entities, (2) it is nonsymmetrical, and (3) it is conditional transitive. Two types of trust relationships are defined: *direct trust* and *recommender trust*. The model categorizes a trust relationship between entities in terms of different interactions. Trust in one category is independent of trust in other categories. For instance, A needs a car service from D who is unknown to A. A trusts B's recommendation and asks him about D in car service category. B does not know D either so it asks C who has D's trust information in terms of car service. C sends D's trust value to A. In this way a recommendation path is formed. When multiple recommendation paths exist, the final trust is an average of all the paths.

However, the Web of trust model has several drawbacks:

- It does not explicitly address false recommendations, which significantly influence the final trust value;

- It does not provide efficient mechanisms to update trust values, which may change frequently in dynamical and distributed environments.

- It is nontrivial to explore a recommendation path in large-scale social networks.

Yao et al. [289] improved the accuracy of trust inference by integrating transitivity, multi-aspect and prior knowledge. A multi-aspect trust model, which treats the trust problem as a recommendation problem, was proposed. Specifically, the model directly characterizes a latent factor vector for each trustor and trustee from the locally-generated trust relationships:

$$\min_{\mathbf{F},\mathbf{G}} \sum_{(i,j \in \mathcal{K})} (\mathbf{T}_{i,j} - \mathbf{F}(i,:)\mathbf{G}(j,:))^2 + \lambda \parallel \mathbf{F} \parallel^2 + \lambda \parallel \mathbf{G} \parallel^2, \qquad (1.2)$$

where $\mathbf{T}_{i,j}$ indicates the trust relationship between user i and user j, which is typically a numeric rating. $\mathbf{F}(i,:)$ and $\mathbf{G}(j,:)$ represent the latent factors of user i and j respectively. $\parallel \mathbf{F} \parallel^2$ and $\parallel \mathbf{G} \parallel^2$ are regularization terms for avoiding overfitting where the parameter λ controls the extent of regularization. The unknown trust ratings can be estimated based on the latent factors of the corresponding users.

The trust transitivity is considered from four aspects: (1) Direct propagation, i.e., $\mathbf{T}^t(i,j)$ measures the direct trust propagation from user i to user j after t steps. (2) Transpose trust, which indicates that user j's trust of user i can cause some level of trust in the opposite direction. (3) Co-citation, which means if two users i and j are both trusted by another user k, then i and j may trust each other to certain extent. (4) Trust coupling, which means if two users i and j both trust another user k, they may trust each other. These four aspects are integrated into a propagation vector $\mathbf{z}_{i,j}$ for the user pair (i,j). And the trust score can be predicted by $\hat{\mathbf{T}}(i,j) = \mathbf{F}(i,:)\mathbf{G}(j,:)' + \alpha[\mu, \mathbf{x}_i, \mathbf{y}_j] + \beta\mathbf{z}_{i,j}$, where μ, $\mathbf{x}_i$ and $\mathbf{y}_j$ represent global bias, trustor bias and trustee bias respectively. α and β are the corresponding weights for prior knowledge (i.e., the bias) and trust transitivity.

1.3.1.4   Iterative Methods

Another class of methods computes trust through transitive iteration. An representative trust model using such a computation method is Eigen-Trust [119], which is a reputation system developed for Peer-to-Peer networks. EigenTrust tries to fulfill the goals of *self-policing, no profit for newcomers, anonymity-maintaining, minimal overhead* and *being robust to malicious collectives of peers*.

EigenTrust calculates a global reputation for each peer in the network based on the local opinions of all other peers. Every peer i stores for every interaction partner j a trust value s_{ij}, representing the experience it has gained. s_{ij} is normalized locally (denoted by c_{ij}) to avoid the arbitrary values assigned by malicious peers. Peer i then asks its acquaintance what he thinks about other peers and receives the normalized trust values. All the normalized values are stored in a matrix. From this matrix, peer i can retrieve any peer j's global reputation t_{ij} by calculating

$$t_{ij} = \sum_{k=0} c_{ik} \cdot c_{kj} \qquad (1.3)$$

So peer's global reputation is the weighted sum of all other peers' opinions where the weight factor is global reputation of the opinion reporter.

Let C be the matrix $[c_{ij}]$, define $\vec{t_i}$ to be $\forall_k t_{ik}$ and $\vec{c_i}$ to be $\forall_k c_{ik}$. Then, the global reputation is $\vec{t_i} = (C^T)\vec{c_i}$. To broaden the view further, a peer will ask its friends' friends' opinions: $((C^T)(C^T))$. Increasing n in $\vec{t_i} = (C^T)^n\vec{c_i}$ continues to broaden the view. Finally, $\vec{t_i}$ will converge to the same vector (i.e., all peers will converge to the same values representing global reputation).

EigenTrust uses distributed hash table (DHT) to store and lookup global reputation scores. A score manager is responsible for calculating, storing and communicating reputation scores for the identities who fall in the score manager's responsibility range in the DHT. The use of multiple hash functions allows multiple score managers for each host to provide redundancy.

Following EigenTrust, some other works (e.g., GossipTrust [300]) relying on a similar method are proposed, to look at different aspects of the problem. For instance, PowerTrust [299] aims to fulfill the key design issues of *high accuracy, fast convergence speed, low overhead, being adaptive to peer dynamics, being robust to malicious* peers and *scalability* for trusted Peer-to-Peer computing. The authors proposed a Trust Overlay Network (TON) built on top of a P2P network, where any pair of interacting peers are connected with a feedback score. When a new interaction occurs between a pair of peers, the feedback score is updated accordingly. Each peer i stores local reputation scores of all the peers with whom it has interacted. Let s_{ij} be the raw trust score i assigns to j, the corresponding local reputation score r_{ij} is normalized as $s_{ij}/\sum_k s_{ik}$. All the local reputation scores are stored in a matrix $R = (r_{ij})$.

The global reputation score v_i of peer i is computed based on other peers' local reputation scores. All the v_i are stored in a normalized reputation vector $V = (v_i)$, where $\sum_i v_i = 1$. V is updated recursively by

$$V_{(t+1)} = R^T \cdot V_{(t)} \qquad (1.4)$$

Computation reaches convergence when $|V_{(t+1)} - V_{(t)}| < \epsilon$, where ϵ is the error threshold. This process is very similar to EigenTrust [119]. To improve aggregation efficiency, a *look-ahead random walk* (LRW) strategy was proposed where each peer combines its local reputation scores and its neighbors' first-hand ones.

Update of reputation follows a random walk through a TON. A greedy factor is defined as the probability of jumping directly to the power nodes. So power nodes are updated more often to avoid one of them starting to subvert the system.

1.3.2 Machine Learning for Trust Modeling

In the previous subsection, we briefly review the traditional trust models from the perspectives of information sources (e.g., direct/indirect experience, stereotypical information) and calculation methods (e.g., Web of trust). In this subsection, we emphasize the recently emerged machine learning-based trust modeling. In Section 1.3.2.1, we introduce some fundamentals of machine learning, based on which we discuss novel machine learning-based trust modeling in Section 1.3.2.2.

1.3.2.1 A Little Bit about Machine Learning

Machine learning, an outgrowth of the intersection of computer science and statistics, aims to automatically learn to recognize complex patterns and make intelligent decisions based on existing datasets. The principle of machine learning is described as follows: Suppose that there is a dataset $\mathbf{X} = (x_1, x_2, ..., x_n)$

and a function (of $\mathbf{X}$) f, the purpose of learning is to get what the function f is, such that the decision (e.g., classification) can be made based on $f(\mathbf{X})$. The hypothesis about the function (of $\mathbf{X}$) to be learned is denoted by h. We assume that h is selected from a class of functions $\mathcal{H}$, and f also belongs to (a subset of) this class. The selection of function estimator h is determined based on the training data $\mathbf{X}$.

There are two main types of learning (i.e., learning the function f). One is *supervised learning* where we know the values of f for a subset of data samples in $\mathbf{X}$. Based on these observations, if the hypothesis h can be determined to approximate f, such h could be a good estimate for f. The other type of learning is *unsupervised learning* where the function f's values are not observed. We refer readers to classic machine learning textbooks such as [95, 193] to get more details of machine learning.

Over the past 50 years of research, machine learning has been successfully applied to many application scenarios. In particular, with the advent of the age of Big Data, as well as the development of distributed and paralleling computing, machine learning has become a promising tool to solve large-scale, complex, real-world problems. For instance, most existing commercial *speech recognition* systems use machine learning (e.g., Hidden Markov Model) to learn a speakers' voice, based on which to fine tune the recognition; modern natural language processing (NLP) algorithms are based on machine learning, especially statistical learning to learn rules by analyzing large text corpora.

In the area of trust research, machine learning also plays an important role. For instance, traditional beta and Dirichlet distribution-based reputation systems employ Bayesian learning to compute and update trust values based on agents' past interaction behavior. Recently, the increasing amount of data as well as the rich metadata brought by large-scale Web applications (e.g., social media, e-commerce, recommender systems) has lead to a new trend of applying formerly unutilized machine learning methodologies, such as supervised learning, to more precisely model trust. In the next subsection, we elaborate how the trust modeling problem is translated into a (supervised) learning problem and discuss the state-of-the-art machine learning-based trust models that tackle some open questions in trust research.

1.3.2.2   Machine Learning for Trust

Imagine that in a system, each agent may have a set of past transactions with other agents, we argue that by investigating useful features (e.g., contextual information that is associated with agents and transactions) that are capable of distinguishing successful transactions from unsuccessful ones, sophisticated machine learning algorithms can be applied to analyze past transactions. If these algorithms manage to model efficiently what a successful (or unsuccessful) transaction is, we can then use this to predict trustworthiness of a potential transaction. In this subsection, we discuss how machine learning can be applied to trust modeling to handle two open questions in trust research: (1) estimating initial trustworthiness of the target agent in the absence of information about its past behavior (direct and/or indirect experience), and (2) capturing the target agent's dynamic behavior in different interactions. The application of machine learning for other trust research questions, as well as its effectiveness in different application scenarios, is discussed in other chapters of this book.

Initial Trust Estimation

When the information about the target agent's past behavior is not available (i.e., cold start), traditional reputation based approaches cannot be directly applied to assess trust. As mentioned in the earlier section, stereotyping based models such as StereoTrust [150] have advanced initial trust estimation by leveraging the trustor's past experience with other relevant agents. However, StereoTrust has its own shortcomings. For instance, it cannot tell which stereotypes are more important than others; the transaction amounts-based weight determination for stereotypes combination is straightforward, intuitive and heuristic. In order to further improve stereotyping-based trust models, machine learning algorithms are applied to more efficiently learn the trustor's local knowledge.

MetaTrust [152, 153] is a generic machine learning framework for identifying relevant features to determine trust. Specifically, a trustor uses its own previous transactions (with other agents) to build a knowledge base, and utilizes this to assess trustworthiness of a potential transaction based on associated features, which are capable of distinguishing successful transactions from unsuccessful ones. These features are harnessed using appropriate machine learning techniques to extract relationships between the potential transaction and previous transactions. MetaTrust is generic in the sense that various machine learning algorithms can be integrated, demonstrating that trustworthiness can be efficiently learned.[3] This work uses two common but effective machine learning algorithms — linear discriminant analysis (LDA) and decision tree (DT) — as the case studies to demonstrate how MetaTrust works.

Specifically, the trustor's past interactions (with other relevant agents) are described/characterized by a set of features. Without loss of generality, two classes are assumed: successful and unsuccessful interactions. Note that feature selection is application dependant, and all the interactions have the same feature set. For instance, in an online auction site, such features may be the price or the category of the item, or the number of items already sold by the seller (see Table 1.1 as an example in Allegro (http://allegro.pl/)). For LDA, the trustor divides his historical interactions into two groups: successful and unsuccessful. He then performs LDA on these two groups to obtain a linear classifier that allows him to estimate whether the potential interaction is likely to get classified in the successful group. For decision tree algorithms, a

---

[3]That does not mean that machine learning can always be used, since its usage still depends on the availability of appropriate information (which happens to be somewhat different from the kind of information needed in traditional trust models).

TABLE 1.1: Structure of Local Knowledge Repository

Tran. ID	Transaction partner	Outcome	Price of the item	♯ of items already sold	...
Θ_1	a_1	successful	6.99	891	...
Θ_2	a_2	unsuccessful	9.99	337	...
Θ_3	a_3	unsuccessful	12.99	120	...
......					

tree is first constructed based the training data. Then the algorithm classifies the potential transaction by starting from the root of the tree and moving (down) until a leaf node, i.e., get classified. Similar to MetaTrust, Burnett et al. used the M5 model tree learning algorithm to learn stereotypes [28].

With sufficient local knowledge and suitable machine learning algorithms, the trustor is able to reliably predict the trustworthiness of a potential transaction. However, when local knowledge is insufficient, machine learning algorithms will perform poorly. To address this issue, a local knowledge sharing overlay network (LKSON) is constructed such that agents are able to share their local information. Different from traditional trust mechanisms where the feedback of the specific agent are shared, in this work, agents only exchange intermediate machine learning algorithm results. Such a strategy has several advantages: (1) the shared information is only the intermediate result of an algorithm so it is not easy to dig out agent's privacy, i.e., identification; (2) since the information provider does not know whom the trustor is evaluating as well as the trustor's local knowledge, it is difficult to send fake information to promote or to bad-mouth some specific agent; (3) a lot of computation duplication is avoided.

From a different perspective, Tang et al. [257] addressed the issue of initial trust assessment using the homophily effect. Homophily suggests that similar users are more likely to establish trust relations. For instance, people with similar tastes about fiction movies tend to trust each other. This work employs low-rank matrix factorization to study trust relations. Specifically, let $\mathbf{u} = \{u_1, u_2, ..., u_n\}$ denote the set of n users. $\mathbf{G} \in \mathbb{R}^{n \times n}$ is a trust relation matrix where $\mathbf{G}(i,j) = 1$ if u_i trusts u_j, and $\mathbf{G}(i,j) = 0$ means no trust relation between u_i and u_j is observed. By performing matrix factorization (i.e., optimizing Equation 1.5), $\mathbf{u}$ is represented by a low-rank, user specific matrix $\mathbf{U} \in \mathbb{R}^{n \times d}$, where $d \ll n$ is the dimensionality of latent factor vector of individual users:

$$L = \min_{\mathbf{U}, \mathbf{V}} \| \mathbf{G} - \mathbf{U} \mathbf{V} \mathbf{U}^T \|_F^2, \tag{1.5}$$

where $\mathbf{V} \in \mathbb{R}^{d \times d}$ is a matrix that captures correlations among low-rank representations, i.e., $\mathbf{G}(i,j) = \mathbf{U}(i,:) \mathbf{V} \mathbf{U}(j,:)^T$. To incorporate the homophily effect, a homophily regularization term is added to the objective function:

$$L = \min_{\mathbf{U},\mathbf{V}} \parallel \mathbf{G} - \mathbf{U}\mathbf{V}\mathbf{U}^T \parallel_F^2 + \alpha \parallel \mathbf{U} \parallel_F^2 + \beta \parallel \mathbf{V} \parallel_F^2 +$$

$$\lambda \sum_{i=1}^{n} \sum_{j=1}^{n} \zeta(i,j) \parallel \mathbf{U}(i,:) - \mathbf{U}(j,:) \parallel_F^2, \qquad (1.6)$$

where $\zeta(i,j) \in [0,1]$ is the homophily coefficient between u_i and u_j. The larger $\zeta(i,j)$ is, the more likely a trust relation is established between u_i and u_j. $\zeta(i,j)$ can be calculated by some common measures such as Jaccard's index and Pearson Correlation Coefficient (PCC). Equation 1.6 can be solved by applying an alternative optimization solution [55] where $\mathbf{U}$ and $\mathbf{V}$ are updated alternatingly.

We can observe that the latent representation for u_i is smoothed with other users via the homophily coefficient. So even for inactive users with a few or even without any trust relations, we still can estimate their latent representations via homophily regularization, thus addressing the data sparsity issue.

Dynamic Trust Prediction

In large-scale, open systems such as online social networks, an intelligent agent may vary its behavior in different interactions with different interaction partners to maximize its profits. For instance, in an online auction site, a malicious seller may act honestly in selling cheap items to gather sufficient reputation and then cheat in selling an expensive item. It is thus essential to model an agent's dynamic trust by capturing its dynamic behavior. Some early attempts on this issue extended the popular beta distribution-based trust models by adopting the "forgetting factor" [115, 264]. However, recent studies have showed that these approaches fail to effectively detect dynamic behavior patterns compared to another class of solutions that are based on Hidden Markov Model (HMM) [186].

In Moe, Tavakolifard and Knapskog [184], a trust model for multi-agent systems is developed to help the agent make optimal trust decisions over time in a dynamic environment. The target agent's behavior is predicted according to the HMM trust estimation module following the Q-learning greedy policy. ElSalamouny et al. [61] modeled the real dynamic behavior of an agent by HMMs. They further justified the consistency of the model by measuring the difference between real and estimated predictive probability distributions using relative entropy. The works [185] and [161] demonstrate how HMM-based trust models are applied to distinct application scenarios: routing protocol design in mobile and ad-hoc networks (MANET) and Web service providers selection.

Although HMM is a promising technique to model dynamic trust, most existing models intuitively use the outcomes of the past interactions as the observation sequence. This method is effective when an agent changes its behavior in specific patterns but is not well suited to identify implicit patterns

from the random behaviors. For instance, when most of an agent's past interactions are satisfactory, it is quite challenging to detect its "sudden" behavior
change. To address this issue, a context-aware HMM-based trust model is
proposed [155]. Liu and Datta first argue that an agent's dynamic behavior is correlated with and can be inferred (to certain extent) by interaction
contextual information. Three sources where such contextual information can
be extracted are identified: (1) from properties of the target agent, (2) from
properties of the service/products that are provided by the target agent and
(3) from properties of the target agent and other agents in the network. An
HMM-based model considering such contextual information is then proposed
to capture the dynamic behavior of the target agent.

Specifically, a set of interactions between the trustor and the target agent is
assumed. Each interaction is associated with contextual information, as mentioned earlier. We also assume the outcome of an interaction has l levels of
trust rating: $\mathcal{L} = \{L_1, L_2, ..., L_l\}$ (e.g., bad, medium, good). Each level corresponds to a trust state of the target agent, so an l-state HMM λ is constructed.
The observation of λ is actually contextual information that is associated with
each interaction. Given the model λ and a sequence of observations of past
interactions $F_{m+1} = f_0 f_1 ... f_m$ with the target agent, the trustor is able to
infer trust state s_m (i.e., a certain trust rating level L_i) of the target agent
in the potential interaction, which is expressed by $P(s_m = L_i | \lambda, F_{m+1})$. A
forward algorithm is applied to recursively derive this probability. Finally, the
trustor is able to decide whether to interact with the target agent based on
the trust rating which has the highest probability.

Tang et al. [259], on the other hand, tried to address the issue of dynamic
trust from a very different perspective. Specifically, the authors studied online
trust evolution by exploiting the dynamics of users' preference in online review
sites like Epinions.com. The proposed trust evolution framework eTrust works
under the assumption that trust is strongly correlated with users' preference
similarity in rating systems [303]. That is, the more similar two users are, the
greater the trust between them exists. So from this viewpoint, trust relationships will evolve with the dynamics of users' preferences. For instance, when
a user is interested in buying a smart phone at time t, she is likely to trust
the expert reviewers in the category of "Phones"; while when she switches
her interest to cars at time $t + 1$, she will trust the expert reviewers in the
category of "Cars".

In order to capture users' preference evolution, and hence the dynamic
trust, the rating from user i to item j is predicted by linearly combining a
latent factor model, which encodes users' preference and a neighborhood-based
model, which incorporates trust information:

$$\hat{r}_{u,v} = \alpha \sum q_j p_i(t) + (1 - \alpha) \frac{\sum_{v \in N_i} w_{v,i} q_j r_{v,j}}{\sum_{v \in N_i} w q_j}, \tag{1.7}$$

where q_j is the latent factors of item j, $p_i(t)$ is the latent factors of user i at
time t, α is the weight for the two models, $w_{v,i}$ is the trust strength between

user i and her neighbor v (N_i is the size of user i's neighbors). Furthermore, an exponential time function is applied to decay the influence of a past rating. The objective function of eTrust is defined as the least square of the predicted ratings and the corresponding real ratings. Projected gradient method is used to optimize the objective function.

1.4 Structure of the Book

This section briefly summarizes the remaining chapters, which apply machine learning based computational trust modeling to various application domains to solve specific real-world questions.

Chapter 2 provides a review on the reputation-based trust research in different online communities, including e-commerce (e.g., eBay), search engine (e.g., Google's PageRank), Peer-to-Peer networks, service-oriented environments and social networks (e.g., Facebook), with the emphasis on the machine learning-based online trust evaluation.

In Chapter 3, we look at the problem of whether to believe a retrieved document (i.e., trustworthiness of the document). A framework that is based on a fact-finder algorithm, which iteratively learns the trustworthiness of a piece of information and the information source, is presented. A wealth of background knowledge and contexts such as nationality and temporal-spatial information are considered to improve the fact-finding iteration process for more accurate decision making.

In Chapter 4, we still cover the issue of assessing the credibility/trustworthiness of Web contents but from a different perspective. Specifically, the Web credibility is evaluated by leveraging crowdsourcing. Data collection (e.g., using Amazon Mechanic Turk) is presented followed by a comprehensive analysis such as the influence of user characteristics, bias and controversy on the credibility evaluation. Finally, an credibility classifier based on Web content features is trained to automatically predict the credibility of an unknown Web content.

Chapter 5 first surveys existing techniques such as content-based and collaborative filtering based approaches for recommender systems, and then discusses how trust information can help to improve the quality of recommendations, based on the assumption that people tend to rely on recommendations from their friends and other people they trust, more than those provided by strangers or people they do not trust. Two lines of research, trust-aware memory-based collaborative filtering and trust-aware model-based collaborative filtering (i.e., latent factor model), are elaborated.

Finally, Chapter 6 focuses on the the objectivity of the view of information sources. That is, studying the level of *bias* of information sources can help to derive an unbiased estimation of the object quality and trustworthiness of information. In this chapter, different types of bias are first identified. Then both supervised and unsupervised approaches are discussed for the task of bias detection. Finally, we provide several ways to alleviate the effects of bias.

Chapter 2

Trust in Online Communities

2.1   Introduction

Internet World Stats[1] show that the number of online users worldwide has reached 2.75 billion as of March 2013, accounting for almost 38.8 percent of the global population. The increase has had a huge impact on the growth of online communities such as e-commerce, social networks and content sharing sites, especially in recent years. Trust has become a crucial factor for users who interact online, due to the limited Web interface that does not allow them to judge the trustworthiness of the interacting partner as in a typical face-to-face interaction. This is because online interactions are more impersonal and automated, provide fewer direct sensory cues, have less immediate gratification, entail more legal uncertainties, and present more opportunities for fraud and abuse [96]. This is even more the case with e-commerce and business transactions which deal with monetary value.

Reputation systems (trust models) promote online trust by identifying true reputation scores of entities (products/services/users) based on others' opinions. Such scores help users determine trustworthy interaction partners and engage in successful online business transactions. For example, in e-commerce, reputation systems collect, distribute and aggregate feedback about the past behavior of buyers and sellers in the system. Buyers who have previously bought products from a seller share their experiences, normally in the form of a numerical rating reflecting the level of satisfaction for transactions with the seller. These ratings are aggregated to represent the seller's reputation. The reputation value is then used by other buyers to make decisions on which sell-

---

[1]http://www.internetworldstats.com/emarketing.htm

ers to do business with. Reputation systems are particularly useful for users with no or very little experience in the interaction environment. These systems help people decide whom to trust, encourage trustworthy behavior, and deter participation by those who are unskilled or dishonest. Prominent reputation systems include those in commerce (e.g., eBay, Epinions), search (e.g., PageRank), blogs (e.g., Blogger), peer-to-peer networks (e.g., EigenTrust), etc. The following sections will outline the importance of trust and how trust evaluation is performed using reputation systems in various online communities.

2.2   Trust in E-Commerce Environments

Trust is of prime importance in e-commerce environments, because of the huge impact they create on online transactions. Research shows that 92 percent of people do research before making a purchase,[2] 85 percent of users read online customer reviews and ratings before making purchases and 75 percent say that positive customer reviews make them trust a business more.[3] Trust in the form of ratings and reviews promotes or demotes a product or service. Figure 2.1 shows the rating system in *TripAdvisor.com*, signifying the extent of trust users have on a particular airline.

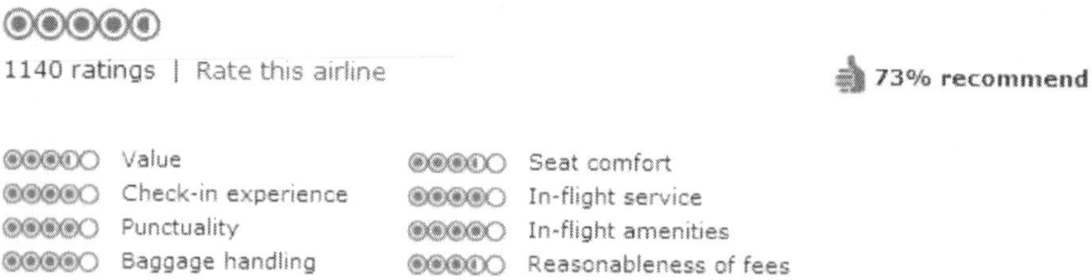

⊚⊚⊚⊚⊙
1140 ratings | Rate this airline   👍 **73% recommend**

⊚⊚⊚⊙○ Value   ⊚⊚⊚⊙○ Seat comfort
⊚⊚⊚⊚○ Check-in experience   ⊚⊚⊚⊚○ In-flight service
⊚⊚⊚⊚○ Punctuality   ⊚⊚⊚⊚○ In-flight amenities
⊚⊚⊚⊚○ Baggage handling   ⊚⊚⊚⊙○ Reasonableness of fees

FIGURE 2.1: Rating system in TripAdvisor.com.

In such open e-commerce environments, it is not easy to establish trust between interacting partners (buyers and sellers) because self-interested sellers can act maliciously by not delivering products with the same quality as promised and buyers can provide misleading opinions [50], to promote some sellers (ballot-stuffing) or to bad-mouth others. A reputation system collects feedback about participants' past behavior and assigns trust scores to each interacting partner. Doing so helps select trustworthy sellers and buyers for successful transactions.

Several trust evaluation schemes for e-commerce systems have been proposed in literature. In the reputation system of eBay, both the buyers and

---

[2]http://reputationx.com/internet-reputation-management/
[3]http://www.marketingcharts.com/

sellers can provide feedback (rating $\in \{1, 0, -1\}$) on each other. A positive rating raises a buyer's or seller's reputation score by 1 point, while a negative rating lowers the reputation score by 1 point. The feedback could also be in the form of text comments. In general, the observed ratings on eBay are positive. Resnick et al. [221] also found that there is a high correlation between buyer and seller ratings, suggesting that there is a degree of reciprocation of positive ratings and retaliation of negative ratings between buyers and sellers on eBay. This is problematic if obtaining honest and fair ratings is a goal, and a possible remedy could be to not let sellers rate buyers. However, as the market has matured, sellers who have accumulated a lot of positive feedbacks were given higher scores, thus accounting for the reliability of the reputation system. Yahoo Auction, Amazon and other auction sites extend eBay's reputation system by using different rating scales or aggregation schemes [219]. Specifically, Amazon allows sellers to be evaluated by buyers on a rating scale of 1-5 stars, as well as to add a text comment. 4-5 stars correspond to positive feedback, 3 stars represent neutral, and 1-2 stars correspond to negative feedback. The overall rating is then calculated according to the average star rating for that particular seller. The Sporas system [294] calculates the trust score based on the ratings of transactions in a recent time period. In this method, the ratings of later transactions are given higher weights as they are more important in trust evaluation. The Histos system proposed in [294] is a more personalized reputation system compared to Sporas. Unlike Sporas, the reputation of a seller in Histos depends on who makes the query and how that person rated other sellers in the online community. Some other fuzzy logic-based reputation models also exist in literature, e.g., Song et al. [248] perform trust evaluation using fuzzy logic and their approach divides sellers into multiple classes of trust ranks (e.g., a 5-star seller or 4-star seller).

Reputation systems in e-commerce play a vital role in distinguishing honest behavior of buyers and sellers from dishonest ones. However, reputation systems have widely become victims to the unfair rating problem, where advisors (i.e., buyers providing feedbacks) provide misleading opinions about sellers, to alter their trust scores. One such reported case occurred when Advertising Standards Agency launched an investigation into the popular review Website *TripAdvisor.com* concerning malicious reviews due to which many small businesses from hospitality industry suffered. These businesses often received defamatory reviews (which sometimes could be a part of a coordinated attack) and almost lost their businesses.[4]

Many trust schemes for multi-agent e-marketplaces have been proposed to deal with the unfair rating problem. The Beta Reputation System (BRS) [115] calculates seller reputation using a probabilistic model based on the beta probability density function, which can be used to represent probability distributions of binary events. The beta distributions are a family of statistical distri-

---

[4]http://www.dailymail.co.uk/

bution functions that are characterized by two parameters α and β. The beta probability density function is defined as

$$beta(p|\alpha, \beta) = \frac{\gamma(\alpha + \beta)}{\gamma(\alpha)\gamma(\beta)} p^{\alpha-1}(1 - p)^{\beta-1} \qquad (2.1)$$

where γ is the gamma function, $p \in [0, 1]$ is a probability variable, and $\alpha, \beta > 0$. To calculate the reputation of a seller, ratings received by the seller are combined by simply aggregating the number of positive ratings (m), signifying that the seller is of high quality, and the number of negative ratings (n), signifying that the seller is of low quality. The reputation of seller s, $R(s)$, is then calculated as the expected value of the beta probability distribution:

$$R(s) = E(p) = \frac{\alpha}{\alpha + \beta} \; ; \; \text{where } \alpha = m + 1 \; ; \beta = n + 1 \qquad (2.2)$$

To handle unfair ratings provided by advisors, Whitby et al. [286] extended BRS to filter out those ratings that are not in the majority amongst other ones by using the iterated filtering approach. Feedback from each advisor to a seller s (both positive and negative ratings) is represented by a beta distribution. If the cumulated reputation of the seller s (obtained using the ratings of all advisors in the market) falls outside the q and $1 - q$ quantile of the beta distribution formed by the advisors' ratings to the seller, then the advisor will be considered dishonest and filtered. However, the iterated filtering approach is only effective when a significant majority of ratings are fair and it filters out the ratings that are not in the majority amongst others. From Figure 2.2, we can see that when the calculated reputation of a seller falls outside the quantile ($q = 0.01$) region (0.01 and 0.99) of the beta distribution formed by a given advisor's ratings to the seller (with $m = 8$ and $n = 2$, respectively), the ratings from that advisor will be considered as unfairly high or unfairly low ratings and filtered.

Teacy et al. [264] propose the TRAVOS model to discount unfair ratings by modeling the trustworthiness of advisors based on their personal experience with the advisors' ratings. This approach is also based on the beta probability density function. It copes with unfair ratings by accomplishing two tasks: 1) It estimates the accuracy of the current feedback (ratings of 1 or 0) provided by the advisor about the seller, by evaluating the buyer's personal experience with the advisor's previous advice. More specifically, it divides the interval of [0, 1] into *bin* number of equal bins. It then finds out all the previous advice provided by the advisor that is similar to the advice being currently given by the advisor. The two pieces of advice are similar if they are within the same bin. The accuracy of the current advice will be the expected value of the beta probability density function representing the amount of successful and unsuccessful interactions between the buyer and the seller, when the buyer follows the previous advice. 2) The approach then adjusts the advisors' ratings according to the obtained accuracy.

The Personalized approach proposed by Zhang and Cohen [298] combines

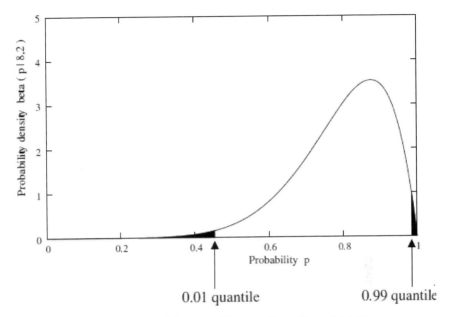

FIGURE 2.2: 1% and 99% quantiles of beta(p|8,2).

buyers' personal experience and the public knowledge held by the system, to model the trustworthiness of the advisors. Private reputation of an advisor a, $R_{pri}(a)$ is calculated (using the beta probability density function) by comparing the advisor's ratings with the buyer's personal ratings regarding the commonly rated sellers. If the ratings are similar, a higher reputation value is achieved. Public reputation $R_{pub}(a)$ is estimated by comparing the advisor's ratings with other advisors' ratings regarding all sellers. The overall reputation of an advisor is then given by

$$R(a) = wR_{pri}(a) + (1 - w)R_{pub}(a) \qquad (2.3)$$

where w is the reliability of the private reputation, calculated based on the minimum number of rating pairs needed to be confident about the private reputation value $R_{pri}(a)$ and the maximal acceptable level of error. A similar approach is adopted to calculate sellers' reputations, i.e., by obtaining a weighted average of the private and public reputation values for sellers.

Several other approaches have also been proposed to deal with unfair ratings. Dellarocas [50] proposed a clustering-based algorithm to separate the advisor's ratings into two clusters (one cluster including lower ratings and another cluster including higher ratings). The ratings in the higher cluster are considered as unfairly high ratings and are discarded. However, this approach cannot effectively handle unfairly low ratings. The iCLUB approach [149] adopts a clustering technique (DBSCAN) to filter out dishonest advisors based on local and global information. Specifically, for a target seller, if advisors' ratings are not in the cluster containing the evaluating buyer's ratings, the advisors are considered to be dishonest. When the buyer has no direct experience (local information) with the target seller, the same process is applied to the non-target sellers to identify the dishonest advisors. Yu and Singh [291] propose a distributed trust model to deal with real ratings. The Dempster-Shafer theory of evidence [127] is used as the underlying computational framework. A real rating is divided into three separate parts by predefined threshold settings and they are allocated into belief, disbelief and uncertainty, respectively. The weighted majority algorithm (WMA) is adopted to adjust the trustworthiness of advisors. If an advisor's opinion of the commonly rated sellers is not same as a buyer's experience, the buyer will decrease its trust value towards the advisor. The BLADE approach of Regan et al. [218] applies Bayesian learning to reinterpret advisors' ratings instead of filtering the unfair ones. The BLADE model allows the buyer to learn other advisors' evaluation functions on different features of the services delivered by sellers by analyzing their ratings. This makes it possible to adjust the advisors' opinions, thereby coping with subjectivity and deception.

In recent times, trust evaluation is frequently based on many criteria, for example, in *TripAdvisor.com*, an airline is rated based on 8 criteria (Figure 2.1): value, check-in experience, punctuality, baggage handling, seat comfort, in-flight service, in-flight amenities and reasonableness of fees. Such detailed trust evaluation benefits users who may have different preferences for the various evaluation criteria. However, it increases the complexity of the trust evaluation engine to compute the multi-criteria trust score and identify the sellers and buyers exhibiting malicious behavior. Some schemes for multi-criteria trust modeling have also been proposed. Griffiths [86] introduced a multi-dimensional trust model tailored to a specific domain with a specific set of criteria. Each criterion is scored as a real number, and heuristics are proposed to update the score based on the buyer's direct experience. The weighted product model [21], which is a standard multi-criteria decision-making technique, is used to combine the different criteria values while calculating the overall reputation of the seller. Each criteria score is raised to the power equivalent to its relative weight according to the evaluating buyer's preferences, while calculating the reputation of seller s, $R(s)$, given by

$$R(s) = \prod_{i=1}^{n} [R_{c_i}(s)]^{w_{c_i}} \tag{2.4}$$

where c is the evaluation criteria, $R_c(s)$ is the reputation score for the seller s on criteria c and w_c is the weight (denoting the buyer's preference) on c. Reece et al. [217] model the seller's reputation by estimating the expected utility of a contract (based on various criteria) which is obtained by determining: 1) the probability that each contract dimension will be successfully fulfilled and 2) the correlations between these estimates. The Dirichlet distribution is used to calculate the probabilities and correlation. If $n_{c_1}, n_{c_2}, n_{c_3}$, etc., represent the number of outcomes for which each of the individual criteria c_1, c_2, c_3, etc., was successfully fulfilled, then in terms of standard Dirichlet parameters, $\alpha_i = n_{c_i} + 1, \alpha_0 = \sum_i \alpha_{c_i} + 2$. The probability of the contract dimension c_i being successfully fulfilled, $p(o_{c_i} = 1)$, and the variance V_{c_i} is given by

$$p(o_{c_i} = 1) = \frac{\alpha_i}{\alpha_0}; \quad V_{c_i} = \frac{\alpha_i(\alpha_0 - \alpha_i)}{\alpha_0^2(1 + \alpha_0)} \tag{2.5}$$

Apart from calculating the estimate of the probability that any contract dimension will be successfully fulfilled, the uncertainty and correlations in these probabilities are also calculated using the covariance matrix. It is found that using the Dirichlet formalism to calculate the multi-criteria trust score is more accurate than using multiple independent Beta distributions for each criterion (ignoring the correlations), because ignoring the correlation between the success probabilities of each criterion will lead to a miscalculation in estimating the uncertainty in the probability of each contract dimension being fulfilled. However, the Reece model evaluates only the reputation of sellers and simply assumes that advisor honesty can be modeled by extending trust models such as TRAVOS [264]. Thus, we can see that the multi-criteria trust schemes, presented above do not address the problem of filtering dishonest advisors in multi-criteria environments to cope with the unfair rating problem. Irissappane et al. [103] propose a biclustering-based approach to detect the malicious behavior of advisors providing misleading opinions specific to multi-criteria environments. Here, each buyer is assigned a set of biclusters, obtained by clustering advisors who are honest to a subset of criteria. Such a mechanism effectively identifies dishonest advisors, who provide honest ratings to some criteria, while acting maliciously on others.

2.3 Trust in Search Engines

Given a query, a search engine identifies the relevant pages on the Web and presents the users with the links to such pages, typically in batches of 10-20 links. Once the users see relevant links, they may click on one or more links in order to visit the corresponding pages. For many commercial Websites, an increase in search engine referrals translates to an increase in sales, revenue and, one hopes, profits.

The early search engines such as Altavista simply presented every Webpage that matched the key words entered by the user, which often resulted in too many and irrelevant pages being listed in the search results. This is because some Web masters may promote Websites in a spam-like fashion by filling Web pages with large amounts of commonly used search key words as invisible text or as meta-data in order for the page to have a high probability of being picked up by a search engine, no matter what the user searched for. To address this problem, current Web search engines use link-based reputation systems (e.g., PageRank) to measure the importance of Web pages and rank them in the order of their reputation scores.

FIGURE 2.3: PageRank: The size of each face is proportional to the total size of the other faces which are pointing to it.

PageRank proposed by Page et al. [200] is a widely used scheme by Google to rank the best search results based on a page's reputation. The reputation of a page (also called PageRank) is based on the number and reputation of other pages which are pointing at it. Figure 2.3 illustrates how PageRank works. Here, the size of each face represents the reputation of a page[5] and is proportional to the total reputation of the other pages pointing to it. In fact, this can be described as a trust scheme, because the collection of hyperlinks to a given page can be seen as public information that can be combined to derive a reputation score. PageRank applies the principle of trust transitivity to the extreme because rank values can flow through looped or arbitrarily long hyperlink chains. If P is a set of hyperlinked pages containing pages u and v, $N(u)$ is the set of Web pages pointing to u, $\widetilde{N}(v)$ is the set of Web pages that v points to and E is some vector over P corresponding to the source of the rank, then the PageRank of u is given by [117, 200],

$$R(u) = cE(u) + c \sum_{v \in N(u)} \frac{R(v)}{|\widetilde{N}(v)|} \qquad (2.6)$$

where c is chosen such that $\sum_{u \in P} R(u) = 1$. The term $cE(u)$ gives the rank based on the initial rank and the term $c\sum_{v \in N(u)} \frac{R(v)}{|\widetilde{N}(v)|}$ gives the rank as a function of the hyperlinks pointing to u.

PageRank has reduced the problem of spam-like pages to a certain extent

---

[5]http://en.wikipedia.org/wiki/PageRank

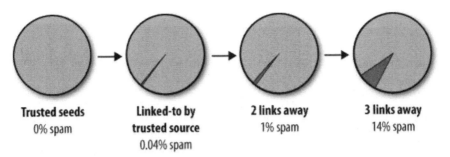

Trusted seeds   **Linked-to by**   **2 links away**   **3 links away**
0% spam   **trusted source**   1% spam   14% spam
                   0.04% spam

FIGURE 2.4: TrustRank.

because a high reputation is also needed in addition to matching key words, in order for a page to be presented to the user while displaying the search results. However, as we know, the common problem in reputation systems is manipulation; strategic users may arrange links attempting to boost their own reputation scores. On the Web, this phenomenon is called link spam, and is usually targeted at PageRank. Though Google's PageRank can deal with this issue to some extent, users can still manage to obtain in-links to boost their own PageRank and can also achieve this goal by carefully placing out-links. There are also certain works which specifically address the issue of link spam. TrustRank [93] addresses the problem of link spam by exploiting the intuition that good pages, i.e., those of high quality, are very unlikely to point to spam pages or pages of low quality. TrustRank propagates trust from the seed set of good pages, recursively to the outgoing links. The trust value is reduced as one moves further and further away from the good seed pages as shown in Figure 2.4.[6] The BadRank algorithm, SpamRank algorithm and Anti-Trust Rank algorithm also deal with this issue [126].

2.4   Trust in P2P Information Sharing Networks

In Peer-to-Peer (P2P) networks, peers communicate directly with each other to exchange information and share files [175] in a decentralized manner (Figure 2.5). All peers are both consumers and providers of resources and can access each other directly without intermediary peers [280]. In an open P2P system, peers often have to interact with unknown peers and need to manage the risks involved in these interactions, as some peers might be buggy or malicious and cannot provide services with the quality that they advertise. For example, before downloading a file, a requesting peer should choose from

---

[6]http://programming4.us/website/1596.aspx

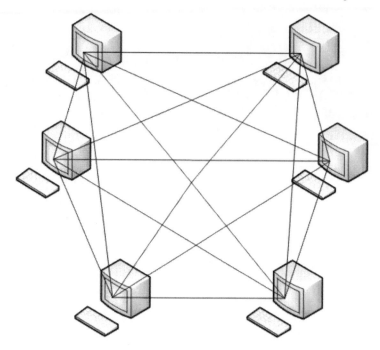

FIGURE 2.5: Peer to Peer (P2P) networks.

a given list of peers that can provide the requested file. The requesting peer then has to check the downloaded file for any malicious content and whether it actually corresponds to the requested file (i.e., the requested content). If the file is corrupted, the requester needs to download the file again. In traditional P2P systems, little information is given to the peers to help in the selection process. Since there is no centralized node to serve as an authority to monitor and punish the peers that behave badly, malicious peers have an incentive to provide poor quality services for their benefit because they can get away with it. Therefore, P2P systems are highly vulnerable to various types of attacks (denial-of-service attacks, etc.). To protect themselves from malicious intentions, requesting peers should be able to identify trustworthy peers for communication, which is quite challenging in such highly dynamic networks.

The issue of trust has been actively studied in P2P information sharing networks (e.g., [46, 119, 280]). Trust in P2P systems allows peers to cooperate and obtain in the long term an increased utility for the participating peers. Here, the requesting peer needs to acquire the trust data of a serving peer (target peer) from other peers, who may have transacted with the serving peer [119, 165, 288]. The computation of the trust level of the serving peer from the collected trust ratings is then performed by the requesting peer rather than a central management server because of the decentralized architecture of the P2P system. However, the major challenge in building such a trust mechanism is to effectively cope with various malicious behavior of peers such as providing fake feedback about other peers. Another challenge is the method of implementation of the trust system in P2P networks. Most existing trust schemes for P2P systems require a central server for storing and distributing the reputation information. Building a decentralized P2P trust management system that is efficient and scalable is quite cumbersome.

EigenTrust [119] is a renowned reputation management algorithm for P2P networks. It adopts a binary rating system, and aims to collect the local trust values of all peers to calculate the global trust value of a given peer. The local trust peer i has on peer j is given by,

$$sat_{ij}^{loc} = sat(i,j) - unsat(i,j) \tag{2.7}$$

where $sat(i,j)$ and $unsat(i,j)$ represent the number of satisfactory and unsatisfactory transactions i has previously had with j. The local trust value is normalized using Equation 2.8, and the global trust is then obtained by aggregating the normalized local trust values from all peers, weighted by their trustworthiness (Equation 2.9).

$$R_{ij}^{loc} = \frac{max(sat_{ij}^{loc}, 0)}{\sum_j max(sat_{ij}^{loc}, 0)} \tag{2.8}$$

$$R_{ik}^{global} = \sum_j R_{ij}^{loc} R_{jk}^{loc} \tag{2.9}$$

The core of the protocol is the normalization process (Equation 2.8), where the trust ratings held by a peer are normalized to have their sum equal to 1. Although it has some interesting properties, this normalization may result in the loss of important trust information. For example, the normalized trust values may not distinguish between a new peer and a peer with whom peer i has had a bad experience. Also, it assumes that there are some peers in the market who are already known to be trustworthy.

Xiong et al. [288] proposed a more efficient solution called PeerTrust to effectively evaluate the trustworthiness of peers and identify various malicious behaviors. They introduce three basic trust parameters (i.e., the feedback that a peer receives from other peers, the total number of transactions that a peer performs and the credibility of the feedback sources) and two adaptive factors (i.e., a transaction context factor to differentiate between transactions and a community context factor to address community specific issues) in computing the trustworthiness of peers. Then, they define some general trust metrics and formulas to aggregate these parameters into a final trust value, given by

$$T(u) = \alpha \times \sum_{i=1}^{I(u)} S(u,i) \times Cr(p(u,i)) \times TF(u,i) + \beta \times CF(u) \tag{2.10}$$

where $T(u)$ is the trust value of peer u, $I(u)$ is the total number of transactions

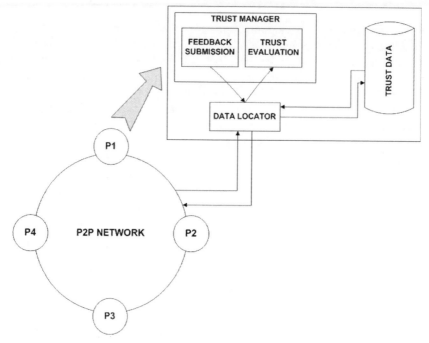

FIGURE 2.6: PeerTrust system architecture.

performed by peer u with all other peers, $S(u,i)$ is the normalized satisfaction u receives in the ith transaction, $p(u,i)$ is the participating peer in the ith transaction, $Cr(p(u,i))$ is the credibility of peer $p(u,i)$, $TF(u,i)$ is the adaptive transaction context factor and $CF(u)$ is the adaptive community context factor. α and β denote the normalized weight factors for the collective evaluation and community context factor. The implementation architecture [288] of PeerTrust is shown in Figure 2.6. There is no central database and the trust data needed to compute the trust measure for peers is stored across the network in a distributed manner. Each peer has a trust manager which 1) provides feedback to the appropriate peers using the data locator and 2) evaluates the trustworthiness of a peer by collecting data from other peers.

Damiani et al. [46] propose an approach for evaluating the reputation of peers through a distributed polling algorithm and the XRep protocol before initiating any download action. This approach adopts a binary rating system and is based on the Gnutella[7] query broadcasting method. The following steps are used in the process: 1) resource searching: initiator p sends a query message for searching resources, peers matching that request respond with a query hit; 2) vote polling: p polls its peers about the reputation of a resource r and the set T of serving peers that offer it, peers wishing to respond send back a poll reply; 3) vote evaluation: p selects a set of reliable voters and contacts

---

[7]http://www.gnutella.com/.

them directly regarding their opinion about r and T; 4) best servant check: p contacts the best serving peer s to check the fact that it exports resource r; 5) resource download: finally p selects s, downloads a resource r, checks its integrity and updates its opinion based on the downloaded resource. Other P2P reputation systems include that of Marti et al. [165], who propose a voting system that collects responses from other peers regarding a target serving peer. The final reputation value is calculated by aggregating the values returned by the responding peers and the requesting peer's experience with the target peer. Zhou et al. [299] explore a power-law distribution in peer feedbacks, and develop a reputation system with a dynamic selection of a small number of power nodes that are the most reputable in the system.

2.5 Trust in Service-Oriented Environments

In service-oriented computing (SOC), service clients interact with service providers for services or transactions. From the point view of service clients the trust status of a service provider is a critical issue to consider, particularly when the service provider is unknown to them. Typically, the trust evaluation is based on the feedback provided by service clients, about the quality of the service providers. In SOC environments, it is more feasible for the central trust management server(s) to compute, manage and distribute the trust values.

In the literature, the issue of trust has also received much attention in the field of service-oriented computing. Vu et al. [276] present a model to evaluate service trust by comparing the advertised service quality with that actually delivered. If the delivered service quality is as good as the advertised service quality, the service is reputable. The model consists of two phases: 1) service discovery: a list of Web services with similar functionalities as required by the user is obtained from a matchmaking framework and 2) service ranking: the user ranks the obtained services based on the predicted quality of service (QoS) values, taking into consideration the explicit quality requirements of users in the queries. For this, user reports on the QoS of all services over time are collected. The predicted QoS values are also based on the quality promised by the service providers, while still considering trust and reputation issues. Wang et al. [279] propose a fuzzy reputation scheme for trust evaluation in SOC environments. The trust value of a service provider at time t_{k+1} is given by

$$R_{k+1}(s) = \begin{cases} min(1, R_k(s) + \theta \times \Delta), \text{if } \Delta \geq 0 \\ max(0, R_k(s) + \theta \times \Delta), \text{if } \Delta < 0. \end{cases} \qquad (2.11)$$

where $\Delta = R_{k+1}(s) - R_k(s)$ is the difference between the actual rating given to the service provider and the previously predicted trustworthiness, $0 \leq \theta \leq 1$ is the impact factor determining the impact of the recent change Δ on the trust calculation, $\theta = \lambda \times f'(R_k(s))$, where $\lambda > 0$, and $f'(R_k(s))$ is the derivative of the curve function $f(R_k(s))$. The curve function $f(R_k(s))$ depicts the trust evaluation function for the service provider, obtained over a period of time. To more accurately reflect the trust status, rather than using a mere numerical value $R_{k+1}(s)$, 5 fuzzy sets, "very low", "low", "moderate", "high" and "very high", along with their membership functions, are set up to categorize $R_{k+1}(s)$ into trust ranks.

Malik et al. [162] propose a decentralized technique to facilitate trust-oriented selection and composition of Web services. The trust value of a service provider s (based on various evaluation criteria), $R(s)$, is given by

$$R(s) = \frac{\sum_i [\frac{\sum_c (\Phi_c(s,i) \times \Psi_c)}{\sum_c \Psi_c} \times \lambda \times Cr(i)]}{\sum_i Cr(i)} \qquad (2.12)$$

where i represents the other service clients in the market, $\Phi_c(s, i)$ is the rating given by i to s for the evaluation criteria c, Ψ_c is the preference of the client evaluating s for criteria c, λ denotes the reputation fader to give more weight to recent ratings and $Cr(i)$ is the credibility of the client i.

Wang et al. [281] describe a super-agent-based framework for Web service selection, where service clients with more capabilities act as super-agents. These super-agents maintain reputation information of the service providers and share such information with other service clients, that have less capabilities than the super-agents. Also, super-agents maintain communities and build a community-based reputation for a service provider based on the opinions from all community members (service clients in a community) that have similar interests and judgement criteria as the super-agents or the other community members. A reward mechanism is also introduced to create incentives for super-agents to contribute their resources (to maintain reputation and form communities) and provide truthful reputation information.

While most of the works on trust evaluation in SOC have focused on accurately predicting trust scores, Conner et al. [41] present a trust model that allows each service client (with different trust requirements) to use different scoring functions over the same feedback data for customized evaluations. Rather than assuming a single global trust metric as with many existing reputation systems, they allow each service client to use its own trust metrics to meet its local trust requirements. They also propose a novel scheme to cache the calculated trust values based on recent client activity.

2.6 Trust in Social Networks

The proliferation of Web-based social networks has led to new innovations in social networking, particularly by allowing users to describe their relationships beyond a basic connection. A social network is a set of people connected by a set of social relationships such as friendship, co-working or information exchange [78]. People share information, express opinions, exchange ideas, make friends and, therefore, form social networks. Some of the famous social networking sites include Facebook.com, Twitter.com and Linkedin.com. The relationships in Web-based social networks are more complex than social network models traditionally studied in the social sciences because users can make a variety of assertions about their relationships with others. For example, users may state how well they know the person to whom they are connected or how much they trust that person. These expanded relationships mean that analysis of the networks can take the new information into account to discover more about the nature of the relationships between people. Also, lots of companies have launched their social media marketing programs on social networking sites, which usually center on efforts to create content that attracts attention and encourages readers to share it with their social networks. For example, large companies such as Adidas have established their communities on social network sites (e.g., *Facebook.com*). Through these communities, users are encouraged to browse and discuss the product information, which can promote the brand reputation of corresponding companies. However, the continuously growing number of users and amount of information with widely varying quality in social networks have also raised the important concern of trust among users, about whom to trust and which information to trust.

Social networks are mainly represented as connected graphs with (directed/undirected) edges representing human-established trust relations (e.g., friend relations) as shown in Figure 2.7. The computational problem of trust is to determine how much one member in the social network should trust another member to whom he is not directly connected. To solve this issue, trust propagation, during which the trust of a target member can be estimated from the trust of other connected members in the social network, is widely used. There are usually many social trust paths between two members, who are unknown to one another (in large-scale social networks, there could be thousands of social trust paths between members). In addition, some social information, such as social relationships between members and the recommendation roles of members, can have a significant influence on the trust evaluation. Thus, evaluating the trustworthiness of a target member based on all the available social trust paths becomes quite challenging and time-consuming. However, one can also search the optimal path yielding the most reliable trust propagation result from multiple paths. We call this the optimal social trust path selection problem, which is still a challenging research problem in this field.

FIGURE 2.7: A social network: edges represent relationships between the individuals.

In the literature, most of the trust evaluation schemes mainly exploit the social network structure and the social interactions between members [243] to select (optimal) social trust paths and accurately determine the trust scores. Buskens [30] observed that high density in a social network (i.e., high interconnectedness between members) can yield a high level of trustworthiness and that members with high out-degree will have higher levels of trust. Caverlee et al. [35] propose a social trust model that exploits both social relationships and feedbacks for trust evaluation. Members in the social network provide feedback ratings after they have interacted with other members. The trust manager combines these feedback ratings to compute the social trust of the members. The member's feedback is also weighted by his link quality (high link quality indicates more links with members having high trust ratings). Golbeck et al. [82] present trust propagation algorithms based on binary ratings. They mainly consider three main concepts for trust evaluation: transitivity, asymmetry and personalization. To illustrate their trust inference scheme, we will consider a social network as shown in Figure 2.8, in which source A has to infer the trust value of F. The source A will first poll each of the neighbors to which it has given a positive rating. Each trusted neighbor (B in this case) will return its rating for the target F. The source will then average these ratings to obtain the inferred reputation rating of F. The source's other neighbors will use this same process to come up with their reputation ratings for F. If there is a direct edge connecting them to the target, the value of that edge is used; otherwise, the value is inferred. In [94], Hang et al. propose an algebraic approach for propagating trust in social networks, including a concatenation

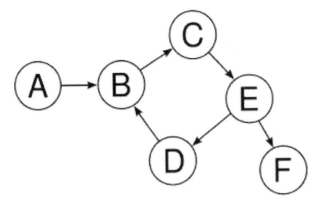

FIGURE 2.8: A social network: node A needs to infer the trust value of F.

operator for the trust aggregation of sequential invocation, an aggregation operator for the trust aggregation of parallel invocation and a selection operator for trust-oriented multiple path selection.

As described above, most existing works for trust inference in social networks use the concept of trust propagation. However, experience with real-world trust systems such as those in Epinions and eBay suggest that distrust is at least as important as trust. To deal with this issue, Guha et al. [87] developed a framework, which uses both trust and distrust propagation, for trust inference in social networks.

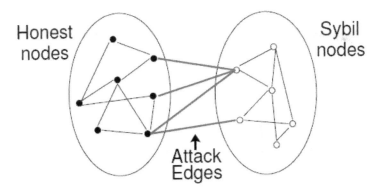

FIGURE 2.9: A social network showing honest and sybil nodes.

Walter et al. [277] identified that network density, similarity of preference between members and sparseness of knowledge about the trustworthiness of recommendations are crucial factors for trust-oriented recommendations in social networks. However, such trust-oriented recommendations can be attacked in various ways, such as sybil attacks, where the attacker creates an unlimited number of false identities to provide feedback and modify the trust score. Yu et al. [292] present SybilGuard, a protocol for limiting the corruptive influences of sybil attacks, which depends on the established trust relationships between members in a social network. Their model is based on the observation that the edges connecting the honest members in the social network and the sybil region (called attack edges) are independent of the number of sybil identities created and are limited by the number of trust relation pairs between the sybil nodes and the honest members. If the malicious members create too many sybil identities, the graph becomes strange, i.e., a small set of edges (attack edges) disconnects a large number of sybil nodes as shown in Figure 2.9.

Zhang et al. [297] propose a scheme to combat sybil attacks in social networks by leveraging on both trust and distrust information. A sybil seed selection algorithm is presented to produce reliable sybil seeds, in combination with current social network-based sybil defense schemes. Moreover, a graph pruning strategy is introduced to reduce the attack ability near honest seeds, by exploiting local structure similarity between neighboring nodes. Finally, a ranking mechanism based on a variant of the PageRank algorithm is presented to combine trust and distrust together, in order to determine the trustworthiness of nodes in the social network and nodes with less trustworthiness score that are more likely to be sybils.

2.7   Discussion

Trust evaluation in online communities has become crucial, mainly because of the risk associated while interacting online. But, it is often difficult to assess the trustworthiness of interaction partners, because computer mediated communication restricts a wide range of normal cues which allow people to easily assess trustworthiness in a physical interaction. The main aim of any reputation system is to distinguish between high and low quality products/services/users by collecting evidence from other members in the online community. However, the nature of the various online communities (e-commerce, search, P2P networks, social networks, etc.) also imposes challenges, with which the reputation system needs to deal.

Online communities widely differ in their structure and content. Thus, reputation systems for the various online communities also differ in their trust evaluation methodology, to be suitable to the application environment. For example, in e-commerce systems buyers and sellers engage in business transactions and the main aim of reputation systems is to select trustworthy sellers by obtaining feedback from trustworthy advisors. Reputation systems for e-commerce environments can operate in a centralized or a decentralized manner, both of which have their advantages and disadvantages. In P2P networks, the main aim of reputation systems is to identify a trustworthy peer. Also, the reputation systems are mainly decentralized in order to suit the P2P network topology. In social networks, the members are represented by nodes of a graph, with edges representing the connectedness, and the reputation systems should find a reliable path between a source node and target node, which are not directly connected. Apart from these differences, the threat models for reputation systems in the various online communities also seem to vary. While in e-commerce systems, unfair rating attacks and malicious seller behavior are of prime importance, in search engines, the reputation systems mainly need to deal with link spams. In P2P systems, again unfair rating behavior is a major concern. In social networks, sybil attacks pose a severe threat, and the reputation systems proposed need to address this issue. Thus, the different nature of each online community demands different trust modeling schemes to address their specific needs and challenges.

Though there is a volume of literature on the theory and applications of reputation systems in online communities, research still needs to focus on the potential fields of improvement, addressing the specific challenges imposed by the online communities and the vulnerabilities of the reputation systems.

Chapter 3

Judging the Veracity of Claims and Reliability of Sources with Fact-Finders

3.1   Introduction

The Information Age has made publishing, distributing and collecting information easier, resulting in the exponential growth of information available to us. Databases were once ledgers written by hand by a single person; today they can be vast stores of data agglomerated from a myriad of disparate sources. The mass media, formerly limited to newspapers and television programs held to strict journalistic standards, has expanded to include collaborative content such as blogs, wikis and message boards. Documents covering nearly every topic abound on the Internet, but the authors are often anonymous and the accuracy uncertain.

To cope with this new abundance, we employ information retrieval to suggest documents and information extraction to tell us what they say, but how can we determine what we should actually *believe*? Not all information sources are equally trustworthy, and simply accepting the majority view often leads to errors: a Google search for "water runs downhill" returns 17.5K documents, while "water runs uphill" yields 116K.

When we consider a collection of data with various authorship, we may view it as a set of information *sources* each making one or more *claims*. Sources often make claims that are contradictory ("Shakespeare was born on April 26, 1564" and "Shakespeare was born on April 23, 1564") and, even in the absence of contradiction, we have no guarantee that the sole presented claim is true. How, then, can we know which claims to believe and which sources to trust? The typical approach is simple: take a vote and choose the claim made by the largest number of sources. However, this implicitly (and implausibly) assumes that all sources are equally trustworthy and, moreover, ignores the wealth of other claims being made by both these and other sources that could inform our belief in the particular claim at hand. For example, if we can ascertain that John's other claims of birthdays for historic figures were correct, his claim about Shakespeare should (ceteris paribus) carry more weight.

A diverse class of algorithms collectively known as *fact-finders* does just this, using the full network of sources and claims to jointly estimate both the trustworthiness of the sources and the believability of the claims. This is useful not just in judging the assertions made by authors in articles, but also in areas such as sensor networks and crowdsourcing. Crowdsourcing of information—where information is polled from a wider population can be done via direct voting, the most famous being reCaptcha [275], which uses humans to solve difficult optical character recognition (OCR) problems as a Turing test and accepts the text for a candidate word image once it has accrued enough votes. Similarly, the ESP Game [274] obtains image labelings, presenting the task as a game. In both cases, the annotators are presented with examples where the labels are already known in an effort to verify their correctness. However, such a coarse approach, where an annotator's trustworthiness is binary, is economically inefficient in systems such as Amazon's Mechanical Turk marketplace where there is a cost for soliciting each vote. Dawid and Skene [47] present a more sophisticated model where each source has a unique, latent confusion matrix that probabilistically determines the likelihood of each possible response to a question, given the truth, with inference performed via Expectation Maximization; this inference can be cast as a fact-finder, though the model is only applicable when the responses (claims) are homogenous across all the questions. Fact-finders have also been directly applied to sensor networks, such as Apollo [139] which received its sensor data via Twitter and inferred the effective speed limit of a road or the true statement about a topic in the varied domains of driving and breaking news, respectively, using an abstract distance metric to cluster similar observations prior to running a standard fact-finder. Such a system is especially invaluable in an emergency, when reports from unreliable first-person sources relayed through social media must be collated and analyzed to discover the moment-to-moment reality.

The basis of each fact-finding algorithm is a two-layer bipartite graph, consisting of a layer of information sources (such as *The New York Times*, Twitter user @truthteller or your friend Bob) with edges connecting them to another layer that consists of the claims they make (Shakespeare's birthday, the height of Everest, etc.) Fact-finders then iteratively calculate the belief in each claim as a function of the trustworthiness of the sources asserting it and the trustworthiness of a source as a function of the belief in the claims it makes. After the stopping criteria is met, the claim taken to be true is simply the one with the highest belief score versus contradicting, mutually exclusive claims (e.g., water runs either uphill or downhill, but not both).

There have been many algorithms for discovering the truth that fit this pattern (e.g., [74, 202, 290]), while others, though intended to solve different problems, have been adapted for this task with minor changes (e.g., Hubs and Authorities [121], originally used for determining the importance of documents). Fact-finders generally have the advantages of being both highly tractable (typically linear time in the number of claims and sources) and easy to describe and implement. Combined with the multitude of extant methods, tuning to find a efficient fact-finder on a given domain is straightforward; this is fortunate since, while clear update rules describe how belief and trustworthiness is determined in each particular iteration, fact-finders are (usually) opaque on the whole—after running to completion there is no concise, precise explanation as to why a particular claim is to be believed over its alternatives. A few fact-finders, such as SimpleLCA [205], are transparent, generative models with explicit semantics, although these are the minority, and, in practice, several opaque fact-finders have, perhaps surprisingly, been shown to work well across a broad range of problems, even without tuning.

Still, since they consider only a bipartite graph of sources and claims, traditional fact-finders do ignore the wealth of background and auxiliary knowledge that are frequently available, including common-sense and specific knowledge about claims, attributes of the sources and the quality of the information extraction. We will also present a framework that addresses this by both generalizing the fact-finding algorithms to admit more informative inputs and enforcing declarative constraints over the iterative process [202, 203]. Each of these two (orthogonal) additions can yield substantially more accurate trust decisions than standard fact-finders on real-world data, both individually and in conjunction. Furthermore, by modeling the user's prior knowledge and beliefs, they allow for *subjective*, rather than objective, truth, often with dramatic practical benefit; it is thus possible, for example, to model the truth of statements such as "travel to Great Britain does not require a visa" relative to the nationality of the user.

3.2 Related Work

An overview of computational trust can be found in [13] and [231]; we next discuss in more detail the foundations of trust for a broader contextualization as well as more concrete work relating to generalized fact-finding and similar problems.

3.2.1 Foundations of Trust

Marsh [164] observes that trust can be global (e.g., eBay's feedback scores), personal (each person has his own trust values) or situational (personal and specific to a context). Fact-finding algorithms are based on global trust, while our framework establishes personal trust by exploiting the user's individual prior knowledge.

Probabilistic logics have been explored as an alternate method of reasoning about trust. Manchala [163] utilizes fuzzy logic [196], an extension of propositional logic permitting [0,1] belief over propositions. Yu and Singh [291] employ Dempster-Shafer theory [239], with belief triples (mass, belief and plausibility) over *sets* of possibilities to permit the modeling of ignorance, while Jøsang, Marsh and Pope [116] uses the related subjective logic [112]. While our belief in a claim is decidedly Bayesian (corresponding to the probability that the claim is true), "unknowns" (discussed later) allow us to reason about ignorance as subjective logic and Dempster-Shafer do, but with less complexity.

Of course, a user's perception of trust cannot be captured entirely by such abstract formalisms. A large amount of work from fields such as human-computer interaction, economics, psychology and other social sciences looks at how trust is created and used by people, which has direct implications in developing an automated system that can take into account the full breadth of relevant information to calculate trust. Much of this research as it pertains to human-computer interaction specifically has been done by Fogg's Persuasive Technologies Lab [14,68–70,267], demonstrating that, besides factors traditionally considered by trust systems such as recommendations and past reliability, humans are superficial, e.g., placing more faith in a site with an .org domain name or a modern, sophisticated design. We note that, though stereotypes, these factors also make useful priors, as .org Websites are often nonprofits that may be more truthful than a Website trying to sell a product, and a sophisticated design implies a high degree of investment by the Website creator who presumably has much more to lose if he violates his visitors' trust than someone whose Website appears to have been prepared in less than an hour. Gil and Artz [79] similarly identify 19 factors that influence trust in the context of the semantic Web, including user expertise (prior knowledge), the popularity of a resource (the wisdom of crowds), appearance (superficial features) and recency (on the basis that more recent information is more likely to be up-to-date and correct). These hitherto exotic factors in trust decisions help motivate our trust framework's ability to incorporate a broad spectrum of prior knowledge, thus permitting them to be leveraged in the context of fact-finding.

3.2.2 Consistency in Information Extraction

Some information extraction systems have confronted the problem of "noise" in their extractions, using various mechanisms to increase the consistency and accuracy of the results. Here, the veracity of the documents is not questioned (explicitly or at all), but rather errors (presumably made by the information extraction itself) are to be corrected and the content of the documents is to be accurately parsed. Such systems can be divided into those that are "local", viewing each document independently, and those that attempt to maintain a global consistency across documents.

3.2.2.1 Local Consistency

Constrained conditional models (CCMs) [36] enforce (soft) constraints across the extracted fields. For example, one constraint might specify that an apartment listing contain only a single price. These constraints do not cross documents and, therefore, cannot leverage redundancy, as they might to identify and resolve inconsistency in the reported prices in two listings for the same apartment, but they can enforce rules within a document, e.g., the prior knowledge that a one-bedroom apartment in San Francisco costs at least $1000.

3.2.2.2 Global Consistency

In Poon and Domingos [211], a Markov Logic Network joint model for both extracting citation fields and identifying coreferential citations is proposed; here, consistency in the information extraction is enforced as a consequence of the (weighted) first-order logic rules; e.g., if a trigram seems likely to start a field in one citation, it probably also starts a field in another citation held to be coreferential. We can consider such constraints to be global is the sense that they cross documents (citations), but such cross-document information could be more accurately regarded as nonbinding hints rather than constraints, and there is no attempt to identify, for example, the true authors of a given work or its canonical title.

Never Ending Language Learning (NELL) [32] is a system for large-scale, ongoing open information extraction, combining a number of subcomponents that infer categories and relations for noun phrases. NELL also induces rules, together with previously inferences, to broaden and (hopefully) improve future extractions (in practice, without human intervention, accuracy declines with time, as might be expected since the "easiest" facts tend to be extracted first). Consistency is achieved by (ontology-based) mutual exclusion and relation-type checking; this allows NELL to know, for example, that an entity cannot be both an actor and a city. Once there exists a preponderance of evidence for a particular candidate fact (i.e., the subcomponents collectively agree or one subcomponent strongly believes *and* the candidate does not conflict with already-believed knowledge), it is permanently added to the knowledge base: NELL does not (yet) revisit previous decisions.

SOFIE [253] enforces consistency with rules which are grounded as propositional logic, and then approximately maximizes the (weighted) number of satisfied clauses. A rule might specify, for example, that one must die within 100 years of birth (note that, as the clauses are weighted, violations are penalized but not prohibited). This is similar to the constraints used by constrained conditional models except that they enforce the consistency of knowledge across documents.

3.2.3   Source Dependence

AccuVote [57,58] is a fact-finding variant that is particularly interesting as it attempts to compute source dependence (where one source copies another) and gives greater credence to more "independent" sources. The idea is that independent confirmation, where multiple sources reach the same conclusion on their own, is much more compelling than a set of dependent sources parroting each other's assertions. Ideally in such cases the copying source would verify the assertions of the copied source before repeating them, but this is often not done in practice. On the benign end of the spectrum, a blog might repost a news story while presuming its veracity with no further consideration. More maliciously, a concerted action across multiple sources is sometimes used to spread misinformation online, e.g., spreading takeover rumors to drive up the price of a stock.

3.2.3.1   Comparison to Credibility Analysis

Credibility analysis models the trustworthiness of each source, whereas information extraction consistency checking effectively treats every source as equally reliable; even if a document overwhelmingly consists of "facts" already disbelieved by the SOFIE system, it will still give equal credence to any others therein. The goals are also different: information extraction seeks to determine what things a document *says*, whereas credibility analysis seeks to determine whether those things should be *believed*. Even in systems that enforce world consistency such as SOFIE, there is a presumption that the documents are truthful and that any violation of the constraints is due to a shortcoming of the model (or the constraints themselves). If two sets of documents simply disagree on a claim, however, the results will be similar to voting (whichever option is asserted most will be believed, or, in the case of NELL, whichever is asserted first!).

TABLE 3.1: Symbols and Their Descriptions

Symbol	Meaning
s	An information source
c	A claim
S	The set of all sources
C_s	The set of claims asserted by $s \in S$
S_c	The set of sources asserting $c \in C$
$C = \bigcup_{s \in S} C_s$	The set of all claims
$M_c \subseteq C$	The *mutual exclusion set* of c
$T^i(s)$	Trustworthiness of source s in iteration i
$B^i(c)$	Belief in claim c in iteration i

3.2.4 Comparison to Other Trust Mechanisms

Reputation-based systems and trust metrics determine trust among peers, with each peer providing recommendations (or disapprovals) for other peers; this may be implicit as in PageRank [22], where the "recommendation" is in the form of a link, or explicitly, as in Advogato [142]. Reputation algorithms thus tend to focus on the transitivity of these recommendations, whereas fact-finders specify the relationship between sources and claims and derive their graph structure from corpora. Wikitrust [9,10] and Zeng et al. [296] are similarly "content-based" and corpus-driven, but these approaches are specialized to Wikis and lack the broader applicability of fact-finders. Lastly, data fusion systems address conflicting claims within a database (e.g., [16] and [44,45]) by examining the provenance (chain of custody) of the data–data that has passed through the hands of trusted agents is more believable than data that has been filtered through one or more less trustworthy agents; in most domains, however, we know only the immediate source of a claim (who said it to us) and not the full provenance, limiting the utility of these approaches.

3.3 Fact-Finding

Before we discuss generalized fact-finding, we'll first formalize the standard fact-finding model. We have a set of sources S, a set of claims C, the claims C_s asserted by each source $s \in S$ and the set of sources S_c asserting each claim $c \in C$. The sources and claims can be viewed as a bipartite graph, where an edge exists between each s and c if $c \in C_s$. In each iteration i, we estimate the trustworthiness $T^i(s)$ of each source s given $B^{i-1}(C_s)$, the belief in the claims it asserts, and the belief $B^i(c)$ in each claim c given $T^i(S_c)$, the trustworthiness of the sources asserting it, continuing until convergence or

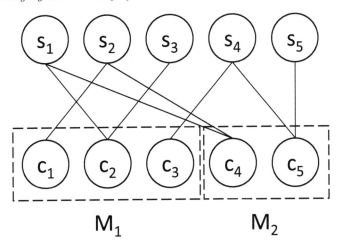

FIGURE 3.1: A small fact-finding problem with five sources and five claims in two mutual exclusion sets, M_1 and M_2.

a stop condition. Note that "trustworthiness" and "belief" as used within a fact-finding algorithm typically do not have well-defined semantics (e.g., they are not $[0, 1]$ probabilities). An initial set of beliefs, $B^0(C)$, serves as priors for each algorithm; these are detailed in the next section. Notice that *any* fact-finder can be specified with just three details: a trustworthiness function $T(s)$, a belief function $B(c)$, and the set of priors $B^0(C)$.

The *mutual exclusion set* $M_c \subseteq C$ is the set of claims that are mutually exclusive to one another (e.g., putative Obama birthplaces) to which c belongs; if c is not mutually exclusive to any other claims, $M_c = \{c\}$. For each mutual exclusion set M containing true claim $\bar{c}$, the goal of the fact-finder is to ensure $\operatorname{argmax}_{c \in M_c} B^f(c) = \bar{c}$ at the final iteration f; the reported accuracies in the empirical results presented later are thus the percentage of mutual exclusion sets we correctly predict over, discounting cases where this is trivial ($|M| = 1$) or no correct claim is present ($\bar{c} \notin M$).

3.3.1 Priors

Except for the 3-Estimates algorithm (where the priors are dictated by the algorithm itself), every fact-finder requires priors for $B^0(C)$. We'll draw from these three: $B^0_{voted}(c) = |S_c| / \sum_{d \in M_c} |S_d|$, $B^0_{uniform}(c) = 1/|M_c|$ and $B^0_{fixed}(c) = 0.5$.

3.3.2   Fact-Finding Algorithms

The fact-finding algorithms we will specifically consider in the remainder of this section are Sums (Hubs and Authorities), Average-Log, Investment, and PooledInvestment, TruthFinder, 3-Estimates.

3.3.2.1   Sums (Hubs and Authorities)

Hubs and Authorities [121] gives each page a hub score and an authority score, where its hub score is the sum of the authority of linked pages and its authority is the sum of the hub scores of pages linking to it. This is adapted to fact-finding by viewing sources as hubs (with 0 authority) and claims as authorities (with 0 hub score):

$$T^i(s) = \sum_{c \in C_s} B^{i-1}(c) \qquad\qquad B^i(c) = \sum_{s \in S_c} T^i(s)$$

We normalize to prevent $T^i(s)$ and $B^i(c)$ from growing unbounded (dividing by $\max_s T^i(s)$ and $\max_c B^i(c)$, respectively), a technique also used with the Investment and Average-Log algorithms; this avoids numerical overflow. B^0_{fixed} priors are used.

3.3.2.2   Average-Log

Computing $T(s)$ as an average of belief in its claims overestimates the trustworthiness of a source with relatively few claims; certainly a source with 90% accuracy over a hundred examples is more trustworthy than a source with 90% accuracy over ten. However, summing the belief in claims allows a source with 10% accuracy to obtain a high trustworthiness score by simply making many claims. Average-Log [202] attempts a compromise, while still using Sums' B^i update rule and B^0_{fixed} priors.

$$T^i(s) = \log |C_s| \cdot \frac{\sum_{c \in C_s} B^{i-1}(c)}{|C_s|}$$

3.3.2.3   Investment

In the Investment algorithm [202], sources "invest" their trustworthiness uniformly among their claims. The belief in each claim then grows according to a nonlinear function $\mathcal{G}$, and a source's trustworthiness is calculated as the sum of the beliefs in their claims, weighted by the proportion of trust previously contributed to each (relative to the other investors). Since claims with higher trust sources get higher belief, these claims become relatively more believed and their sources become more trusted. The experimental results we will present use $\mathcal{G}(x) = x^g$ with $g = 1.2$, together with B^0_{voted} priors.

$$T^i(s) = \sum_{c \in C_s} B^{i-1}(c) \cdot \frac{T^{i-1}(s)}{|C_s| \cdot \sum_{r \in S_c} \frac{T^{i-1}(r)}{|C_r|}}$$

$$B^i(c) = \mathcal{G}\left(\sum_{s \in S_c} \frac{T^i(s)}{|C_s|}\right)$$

3.3.2.4   PooledInvestment

As in Investment, in PooledInvestment [202] sources uniformly invest their trustworthiness in claims and obtain corresponding returns, so $T^i(s)$ remains the same, but now after the belief in the claims of mutual exclusion set M have grown according to $\mathcal{G}$, they are linearly scaled such that the total belief of the claims in M remains the same as it was before applying $\mathcal{G}(x) = x^g$, with $g = 1.4$ and $B^0_{uniform}$ priors used in the experiments. Given $H^i(c) = \sum_{s \in S_c} \frac{T^i(s)}{|C_s|}$, we have

$$B^i(c) = H^i(c) \cdot \frac{\mathcal{G}(H^i(c))}{\sum_{d \in M_c} \mathcal{G}(H^i(d))}$$

3.3.2.5   TruthFinder

TruthFinder [290] is pseudoprobabilistic: the basic version of the algorithm below calculates the "probability" of a claim by assuming that each source's trustworthiness is the probability of it being correct and then averages claim beliefs to obtain trustworthiness scores. We also used the "full", more complex TruthFinder, omitted here for brevity. $B^0_{uniform}$ priors are used for both.

$$T^i(s) = \frac{\sum_{c \in C_s} B^{i-1}(c)}{|C_s|}$$

$$B^i(c) = 1 - \prod_{s \in S_c} \left(1 - T^i(s)\right)$$

3.3.2.6   3-Estimates

3-Estimates [74], also omitted for brevity, differs from the other fact-finders by adding a third set of parameters to capture the "difficulty" of a claim, such that correctly asserting a difficult claim confers more trustworthiness than asserting an easy one; knowing the exact population of a city is harder than knowing the population of Mars (presumably 0) and we should not trust a source merely because it provides what is already common knowledge.

3.4   Generalized Constrained Fact-Finding

If one author claims that Mumbai is the largest city in the world, and another claims that it is Seoul, whom should we believe? One or both authors could be intentionally lying, honestly mistaken or, alternatively, of different viewpoints of what constitutes a "city" (the city proper? the metropolitan area?) Even here, truth is not objective: there may be many valid definitions of *city*, but we should believe the claim that accords with our *user's* viewpoint. Rarely is the user's or author's perspective explicit (e.g., an author will not fully elaborate "the largest city by metropolitan area bounded by ..."), but it is often implied (e.g., a user's statement that "I already know the population of city A is X, city B is Y ..." implies that his definition of a city accords with these figures). A standard fact-finder, however, knows nothing about the user's prior belief and viewpoint and presumes instead to find the (frequently non-existent) objective truth that holds for everyone.

Of course, domain knowledge is not limited to specific statements such as "Los Angeles is more populous than Wichita", but also includes common-sense reasoning such as "cities usually grow over time". We may also know something about the information sources (e.g., "John works for the U.S. census"), the source's own certainty in the claim ("I am 60% certain that ..."), the information extraction system's certainty in the claim ("it is 70% certain that John claimed he was 60% certain that ..."), and the similarity between mutually exclusive claims (if John thinks the population of a city is 1,000, he disagrees less with a claim that the population is 1,200 than a claim that it is 2,000).

Motivated by this, we can synthesize a framework that both *generalizes* fact-finding to incorporate source and similarity information [204] and *constrains* it to enforce the user's common-sense and specific declarative knowledge about the claims [202]. As these aspects are orthogonal and complementary, we introduce them separately, experimentally demonstrating their individual contributions to performance before combining them into a single system able to leverage a very broad range of relevant information for making trust decisions while still building upon the diversity and tractability of existing state-of-the-art fact-finding algorithms.

3.5   Generalized Fact-Finding

The key technical idea behind generalized fact-finding is that we can quite elegantly encode the relevant background knowledge and contextual detail by replacing the bipartite graph of standard fact-finders with a new weighted *k*-

partite graph, transitioning from binary assertions to weighted ones ("source s claims c with weight x"), rewriting the fact-finders to take advantage of these weights (discussed next) and adding additional "layers" of nodes to the graph to represent source groups and attributes (as discussed later).

3.5.1 Rewriting Fact-Finders for Assertion Weights

Generalized fact-finders use weighted assertions, where each source s asserts a claim c with weight $\omega(s,c) = [0,1]$. A surprisingly large amount of information, including the uncertainty of the information extractor, the uncertainty of the source itself, the similarity of claims, and the group membership and attributes of sources, can be encoded into this weight (we will present the details and formula for calculating $\omega(s,c)$ in the next section). After calculating the weights $\omega(s,c)$ for all $s \in S$ and $c \in C$, we need to rewrite each fact-finder's $T(s)$, $B(c)$ and $B^0(c)$ functions to use these weights in the generalized fact-finding process by qualifying previously "whole" assertions as "partial", weighted assertions. We start by redefining S_c as $\{s : s \in S, \omega(s,c) > 0\}$, and C_s as $\{c : c \in C, \omega(s,c) > 0\}$. The basic rewriting rules are

- Replace $|S_c|$ with $\sum_{s \in S_c} \omega(s,c)$
- Replace $|C_s|$ with $\sum_{c \in C_s} \omega(s,c)$
- In $T^i(s)$, replace $B^{i-1}(c)$ with $\omega(s,c)B^{i-1}(c)$
- In $B^i(c)$, replace $T^i(s)$ with $\omega(s,c)T^i(s)$

These rules suffice for all the linear fact-finders we have encountered; one, TruthFinder, is instead log-linear, so an exponent rather than a coefficient is applied, but such exceptions are straightforward.

3.5.1.1 Generalized Sums (Hubs and Authorities)

$$T^i(s) = \sum_{c \in C_s} \omega(s,c)B^{i-1}(c) \qquad B^i(c) = \sum_{s \in S_c} \omega(s,c)T^i(s)$$

3.5.1.2 Generalized Average·Log

Average·Log employs the same belief function as Sums, so we list only the trustworthiness function:

$$T^i(s) = \log\left(\sum_{c \in C_s} \omega(s,c)\right) \cdot \frac{\sum_{c \in C_s} \omega(s,c)B^{i-1}(c)}{\sum_{c \in C_s} \omega(s,c)}$$

3.5.1.3 Generalized Investment

The Investment algorithm requires sources to "invest" their trust uniformly in their claims; we rewrite this such that these investments are weighted by ω.

$$T^i(s) = \sum_{c \in C_s} \frac{\omega(s,c)B^{i-1}(c)T^{i-1}(s)}{\sum_{c \in C_s} \omega(s,c) \cdot \sum_{r \in S_c} \frac{\omega(r,c)T^{i-1}(r)}{\sum_{b \in C_r} \omega(r,b)}}$$

$$B^i(c) = \mathcal{G}\left(\sum_{s \in S_c} \frac{\omega(s,c)T^i(s)}{\sum_{c \in C_s} \omega(s,c)} \right)$$

3.5.1.4 Generalized PooledInvestment

PooledInvestment utilizes the same trustworthiness function as Investment, but instead alters the belief function, which we generalize below.

$$H^i(c) = \sum_{s \in S_c} \frac{\omega(s,c)T^i(s)}{\sum_{c \in C_s} \omega(s,c)}$$

$$B^i(c) = H^i(c) \cdot \frac{\mathcal{G}(H^i(c))}{\sum_{d \in M_c} \mathcal{G}(H^i(d))}$$

3.5.1.5 Generalized TruthFinder

TruthFinder [290] has both a "simple" and "complete" version, with the latter making a number of adjustments to the former. We specify only the simple version below, as the modifications to the complete variant are similar. Both models calculate claim belief nonlinearly, and in either case we have the option of using logarithms to obtain a log-linear function. This is what we do in practice, since it avoids underflow in the floating-point variables; for clarity, however, we present the "multiplicative" version below. Note that using $\omega(s,c)$ as an exponent here is equivalent to its use as a coefficient in the log-linear function.

$$T^i(s) = \frac{\sum_{c \in C_s} \omega(s,c)B^{i-1}(c)}{\sum_{c \in C_s} \omega(s,c)}$$

$$B^i(c) = 1 - \prod_{s \in S_c} \left(1 - T^i(s)\right)^{\omega(s,c)}$$

3.5.1.6 Generalized 3-Estimates

3-Estimates [74] incorporates an additional set of parameters to model the "hardness" of each claim (referred to as $\varepsilon(\mathcal{F})$) that can be incorporated into the B and T functions to fit our common model. We omit the full algorithm here for brevity, but generalizing it is quite straightforward—when calculating a summation over sources for a given claim or a summation over claims for a given source, we simply weight each element of the sum by the relevant assertion weight between the particular source and claim in question.

3.5.2 Encoding Information in Weighted Assertions

As previously mentioned, weighted assertions allow us to encode a variety of factors into the model:

- Uncertainty in information extraction: we have a $[0, 1]$ probability that source s asserted claim c.

- Uncertainty of the source: a source may qualify his assertion ("I'm 90% certain that ....").

- Similarity between claims: a source asserting one claim also implicitly asserts (to a degree) similar claims.

- Group membership: the other members of the groups to which a source belongs implicitly support (to a degree) his claims.

We separately calculate ω_u for uncertainty in information in extraction, ω_p for uncertainty expressed by the source, ω_σ for the source's implicit assertion of similar claims and ω_g for a source's implicit assertion of claims made by the other members of the groups to which he belongs. These are orthogonal, allowing us to calculate the final assertion weight $\omega(s, c)$ as $\omega_u(s, c) \times \omega_p(s, c) + \omega_\sigma(s, c) + \omega_g(s, c)$. Here, $\omega_u(s, c) \times \omega_p(s, c)$ can be seen as our expectation of the $[0, 1]$ belief the source s has in claim c given the possibility of an error in information extraction, while $\omega_\sigma(s, c)$ and $\omega_g(s, c)$ redistribute weight based on claim similarity and source group membership, respectively.

3.5.2.1 Uncertainty in Information Extraction

The information extractor may be uncertain whether an assertion occurs in a document due to intrinsic ambiguities in the document or error from the information extractor itself (e.g., an optical character recognition mistake or an unknown verb); in either case, the weight is given by the probability $\omega_u(s, c) = P(s \rightarrow c)$.

3.5.2.2 Uncertainty of the Source

Alternatively, the source himself may be unsure. This may be specific ("I am 60% certain that Obama was born in ...") or vague ("I am pretty sure that ..."); in the latter case, we assume that the information extractor will assign a numerical certainty for us, so that in either event we have $\omega_p(s, c) = P_s(c)$, where $P_s(c)$ is the estimate provided by source s of the probability of claim c.

3.5.2.3   Similarity between Claims

Oftentimes a meaningful similarity function exists among the claims in a mutual exclusion set. For example, when comparing two possible birthdays for Obama, we can calculate their similarity as the inverse of the time between them, e.g., $|days(date1) - days(date2)|^{-1}$ (where $days$ measures the number of days relative to an arbitrary reference date). A source claiming $date1$ then also claims $date2$ with a weight proportional to this degree of similarity, the idea being that while $date2$ is not what he claimed, he will prefer it over other dates that are even *more* dissimilar. Given a $[0, 1]$ similarity function $\sigma(c_1, c_2)$, we can calculate

$$\omega_\sigma(s, c) = \sum_{d \in M_c, d \neq c} \omega_u(s, d)\omega_p(s, d)\sigma(c, d)$$

Notice that a self-consistent source will not assert multiple claims in mutual exclusion set M with $\sum_{c \in M} \omega_u(s, c)\omega_p(s, c) > 1$, so the addition of $\omega_\sigma(s, c)$ to our formula for $\omega(s, c)$ will never result in $\omega(s, c) > 1$; it is possible, however, that $\sum_{c \in M} \omega(s, c) > 1$ for a given source s. One way to avoid this is to redistribute (smooth) weight rather than add it; we introduce the parameter α to control the degree of smoothing and obtain

$$\omega_\sigma^\alpha(s, c) = \sum_{d \in M_c, d \neq c} \left(\frac{\alpha \omega_u(s, d)\omega_p(s, d)\sigma(c, d)}{\sum_{e \in M_d, e \neq d} \sigma(d, e)} \right) \\ - \alpha \omega_u(s, c)\omega_p(s, c)$$

This function ensures that only a portion α of the source's expected belief in the claim, $\omega_u(s, c)\omega_p(s, c)$, is redistributed among other claims in M_c (proportional to their similarity with c), at a cost of $\alpha \omega_u(s, c)\omega_p(s, c)$.

Yin, Yu and Han [290] previously used a form of additive similarity as "Implication" functions in TruthFinder; however, the formalization presented here generalizes this idea and allows us to apply it to other fact-finders as well.

3.5.2.4   Group Membership via Weighted Assertions

Oftentimes a source belongs to one or more groups; for example, a journalist may be a member of professional associations and an employee of one or more publishers. Our assumption is that these groups are *meaningful*, that is, sources belonging to the same group tend to have similar degrees of trustworthiness. A prestigious, well-known group (e.g., the group of administrators in Wikipedia) will presumably have more trustworthy members (in general) than a discredited group (e.g., the group of blocked Wikipedia editors). The approach discussed in this section encodes these groups using ω_g; a more flexible approach, discussed later, alternatively adds additional "layers" of groups and attributes to create a k-partite rather than bipartite graph.

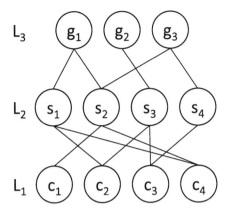

FIGURE 3.2: A fact-finding problem with a single group layer.

Let G_s be the set of groups to which a source s belongs. If a source s and source u are both members of the same group g, we interpret this as an implicit assertion by u in C_s, and by s in C_u—that is, group members mutually assert each others' claims to a degree. We use a redistribution parameter β such that the original weight of a member's assertion is split between the member (proportional to $1 - \beta$) and the other members of the groups to which he belongs (proportional to β). This gives us

$$
\omega_g^\beta(s, c) = \beta \sum_{g \in G_s} \sum_{u \in g} \frac{\omega_u(u, c)\omega_p(u, c) + \omega_\sigma(u, c)}{|G_u| \cdot |G_s| \cdot \sum_{v \in g} |G_v|^{-1}}
$$
$$
- \beta(\omega_u(s, c)\omega_p(s, c) + \omega_\sigma(s, c))
$$

$\sum_{v \in g} |G_v|^{-1}$ in the denominator gives greater credence to "small" groups (where members belonging to many other groups weigh less heavily), with the intuition that smaller groups have more similar members. Note that in the worst case (where all sources belong to a single group and each assert a unique set of k claims), this can effectively create as many as $(k \cdot |S|)^2 - k \cdot |S|$ new assertions, with a corresponding increase in computational cost when running the fact-finder.

3.5.3   Encoding Groups and Attributes as Layers of Graph Nodes

Instead of using weighted assertions, we can add additional "layers" to represent groups and attributes directly. Each node in these layers will represent a group or attribute, with edges linking to its adjoining layers (either the sources or other groups/attributes), creating a k-partite graph (with $k > 3$ used to encode meta-groups and meta-attributes). A standard fact-finder iteratively alternates between calculating the first layer (the claims) and the second layer (the sources), using the B and T functions, respectively. Now we replace these with generic "up" and "down" functions for each layer. For a k-partite graph with layers $L_{1\ldots k}$, we define $U_j^i(L_j)$ over $j = 2 \ldots k$ and $D_j^i(L_j)$ over $j = 1 \ldots k - 1$, with special cases $U_1^i(L_1) = D_1^{i-1}(L_1)$ and $D_k^i(L_k) = U_k^i(L_k)$. The U_j and D_j functions may differ for each layer j, or they may be the same over all layers. In each iteration i, we calculate the values $U_j^i(L_j)$ for layers $j = 2$ to k, and then calculate $D_j^i(L_j)$ for layers $j = k - 1$ to 1. For example, to extend Sums to k layers, we calculate $U_j(e)$ and $D_j(e)$ as follows for $e \in L_j$:

$$U_j^i(e) = \sum_{f \in L_{j-1}} \omega(e, f) U_{j-1}^i(f)$$

$$D_j^i(e) = \sum_{f \in L_{j+1}} \omega(e, f) D_{j+1}^i(f)$$

Where $\omega(e, f) = \omega(f, e)$ is the edge weight between nodes e and f, if e or f is a group or attribute, $\omega(e, f)$ is 1 if e has attribute or group f or vice versa, and 0 otherwise. In many cases, though, we may benefit from using an existing fact-finder over the claim and source layers, while using a different set of functions to mediate the interaction between the source and group/attribute layers. In particular, an information bottleneck often exists when calculating trustworthiness of a source in the "down phase", as it will be wholly dependent upon the trustworthiness of the groups to which it belongs: a source belonging to one overall-mediocre group may make many correct claims, but still be assigned a low trustworthiness score by the D function because of its group membership. This type of problem can be resolved by incorporating both the layer below *and* the layer above in each calculation; for example, for a given $D_j(e)$, we can define $\omega_{children} = \sum_{f \in L_{j-1}} \omega(e, f)$ and $D_j^{smooth}(e) = (1 + \omega_{children})^{-1} D_j(e) + \omega_{children}(1 + \omega_{children})^{-1} U_j(e)$, which returns a mixture of the value derived from e's ancestors, $D_j(e)$, and the value derived from e's descendants, $U_j(e)$, according to the (weighted) number of children e possesses, the intuition being that with more children the trustworthiness computed by $U_j(e)$ is more certain and should be weighted more highly, whereas with fewer children we should depend more upon broad inferences made from the ancestor groups and attributes. We will use $D_j^{smooth}(e)$ in our experiments.

3.5.3.1   Source Domain Expertise

The idea of incorporating the domain expertise of the source into the trust decision has been around at least as far back as Marsh's 1994 thesis [164] and, in the generalized fact-finding framework, we can model it using the same techniques we used to model groups. For example, we expect a plant biologist to be more trustworthy on topics such as photosynthesis and genetic engineering, but less reliable on topics outside his expertise, such as fusion and

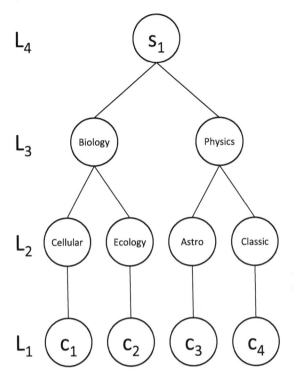

FIGURE 3.3: Use of additional layers to model specialization of the source s_1 into two categories.

computational complexity. Still, these aspects are not entirely separate: if we have two plant biologists A and B, and A gives accurate information about plant biology while B gives inaccurate information, we will tend to assign greater credence to A with respect to other domains (such as physics) as well; that is, we assume A is more "generally trustworthy" overall.

To implement this, we may create additional layers to represent our trustworthiness in a source for various domains. In Figure 3.3, we see that source s_1 has made claims in four different domains: cellular biology, ecology, astronomy, and classical mechanics. Each node shown in layers 2 and 3 is specific to s_1, representing his trustworthiness in that particular field or subfield, such that the sources are actually s_1's "Cellular", "Ecology", "Astro" and "Classic" nodes, which belong to two groups corresponding to s_1's biology and physics trustworthiness, which themselves belong to a metagroup corresponding to s_1's general trustworthiness.

3.5.3.2   Additional Layers versus Weighted Edges

Relative to adding edges to represent groups, expanding our model with additional layers increases the complexity of the algorithm, but prevents the quadratic expansion of the number of edges and the corresponding increase in time complexity. More importantly, though, the flexibility in specifying the U and D functions for the higher layers representing groups and attributes allows us to augment an existing fact-finder to take advantage of them in a highly flexible way.

3.6   Constrained Fact-Finding

In constrained fact-finding, we incorporate rules and constraints to fact-finding. Concretely, to apply the user's specific and common-sense prior knowledge of claims to a fact-finding algorithm, we translate the problem into a linear program (LP). We then iterate the following until convergence or other stopping criteria:

1. Compute $T^i(s)$ for all $s \in S$
2. Compute $B^i(c)$ for all $c \in C$
3. "Correct" the beliefs $B^i(C)$ by applying the linear program

3.6.1   Propositional Linear Programming

To translate prior knowledge into a linear program, we first propositionalize our first-order formulae into propositional logic [229]. For example, assume we know that Tom is older than John and a person has exactly one age $(\exists_{x,y} Age(Tom, x) \wedge Age(John, y) \wedge x > y) \wedge (\forall_{x,y,z} Age(x, y) \wedge y \neq z \Rightarrow \neg Age(x, z))$, and we are considering the following claims: $Age(Tom, 30)$, $Age(Tom, 40)$, $Age(John, 25)$, $Age(John, 35)$. Our propositional clauses (after removing redundancies) are then $Age(Tom, 30) \Rightarrow Age(John, 25) \wedge (Age(Tom, 30) \oplus Age(Tom, 40)) \wedge (Age(John, 25) \oplus Age(John, 35))$.

Each claim c will be represented by a proposition and ultimately a $[0, 1]$ variable in the linear program corresponding informally to $P(c)$.[1] Propositionalized constraints have previously been used with *integer* linear programming (ILP) using binary $\{0, 1\}$ values corresponding to $\{false, true\}$ to find an (exact) consistent truth assignment minimizing some cost and solving a global inference problem, see e.g., Roth and Yih [227, 228]. However, propositional linear programming has two significant advantages:

---

[1]This is a slight mischaracterization, since our linear constraints only *approximate* intersections and unions of events (where each event is "claim c is true"), and we will be satisfying them subject to a linear cost function.

1. ILP is "winner take all", shifting all belief to one claim in each mutual exclusion set (even when other claims are nearly as plausible) and finding the single most believable consistent *binary assignment*; we instead wish to find a *distribution* of belief over the claims that is consistent with our prior knowledge and as close as possible to the distribution produced by the fact-finder.

2. Linear programs can be solved in polynomial time (e.g., by interior point methods [120]), but ILP is NP-hard.

To create our constraints, we first convert our propositional formula into conjunctive normal form. Then, for each disjunctive clause consisting of a set P of positive literals (claims) and a set N of negations of literals, we add the constraint $\sum_{c \in P} c_v + \sum_{c \in N} (1 - c_v) \geq 1$, where c_v denotes the $[0, 1]$ variable corresponding to each c. The left-hand side is the union bound of at least one of the claims being true (or false, in the case of negated literals); if this bound is at least 1, the constraint is satisfied. This optimism can dilute the strength of our constraints by ignoring potential dependence among claims: $x \Rightarrow y$, $x \vee y$ implies y is true, but since we demand only $y_v \geq x_v$ and $x_v + y_v \geq 1$, we accept any value of y_v such that $y_v \geq x_v \geq 1 - y_v$. However, when the claims are mutually exclusive, the union bound is exact; a common constraint is of the form $q \Rightarrow r^1 \vee r^2 \vee \ldots$, where the r literals are mutually exclusive, which translates exactly to $r_v^1 + r_v^2 + \ldots \geq q_v$. Finally, observe that mutual exclusion amongst n claims $c^1, c^2, \ldots, c^n$ can be compactly written as $c_v^1 + c_v^2 + \ldots + c_v^n = 1$.

3.6.2 Cost Function

Having seen how first-order logic can be converted to linear constraints, we now consider the cost function, a distance between the new distribution of belief satisfying our constraints and the original distribution produced by the fact-finder.

First we determine the number of "votes" received by each claim c, computed as $w_c = \omega(B(c))$, which should scale linearly with the certainty of the fact-finder's belief in c. Recall that the semantics of the belief score are particular to the fact-finder, so different fact-finders require different vote functions. TruthFinder has pseudoprobabilistic $[0,1]$ beliefs, so we use $\omega_{inv}(x) = \min((1-x)^{-1}, m_{inv})$ with $m_{inv} = 10^{10}$ limiting the maximum number of votes possible; we assume $1/0 = \infty$. ω_{inv} intuitively scales with "error": a belief of 0.99 receives ten times the votes of 0.9 and has a tenth the error (0.01 vs. 0.1). For the remainder of the fact-finders whose beliefs are already "linear", we use the identity function $\omega_{idn}(x) = x$.

The most obvious choice for the cost function might be to minimize "frustrated votes": $\sum_{c \in C} w_c(1 - c_v)$. Unfortunately, this results in the linear solver generally assigning 1 to the variable in each mutual exclusion set with the most votes and 0 to all others (except when constraints prevent this), shifting all belief to the highest-vote claim and yielding poor performance. Instead, we wish to satisfy the constraints while keeping each c_v close to ω_c/ω_{M_c}, where $\omega_{M_c} = \sum_{d \in M_c} \omega_d$, and thus shift belief among claims as little as possible. We use a weighted Manhattan distance called **VoteDistance**, where the cost for increasing the belief in a claim is proportional to the number of votes against it, and the cost for decreasing belief is proportional to the number of votes for it:

$$ \sum_{c \in C} \max \left(\begin{array}{c} (\omega_{M_c} - \omega_c) \cdot (c_v - \omega_c/\omega_{M_c}), \\ \omega_c \cdot (\omega_c/\omega_{M_c} - c_v) \end{array} \right) $$

The belief distribution found by our linear program will thus be the one that satisfies the constraints while simultaneously minimizing the number of votes frustrated by the change from the original distribution. Note that for any linear expressions e and f, we can implement $\max(e, f)$ in the objective function by replacing it with a new $[-\infty, \infty]$ helper variable x and adding the linear constraints $x \geq e$ and $x \geq f$.

3.6.3   Values → Votes → Belief

Solving the linear program gives us $[0, 1]$ values for each variable c_v, but we need to calculate an updated belief $B(c)$. Pasternack and Roth [202] propose two methods for this:

Vote Conservation: $B(c) = \omega^{-1}(c_v \cdot \omega_{M_c})$
Vote Loss: $B(c) = \omega^{-1}(\min(\omega_c, \ c_v \cdot \omega_{M_c}))$

ω^{-1} is an inverse of the vote function: $\omega_{idn}^{-1}(x) = x$ and $\omega_{inv}^{-1}(x) = 1 - (1 + y)^{-1}$. Vote Conservation reallocates votes such that the total number of votes in each mutual exclusion set, ω_M, remains the same after the redistribution. However, if the constraints force c to lose votes, should we believe the other claims in M_c more? Under Vote Loss, a claim can *only* lose votes, ensuring that if other claims in M_c become less believable, c does not itself become more believable relative to claims in other mutual exclusion sets. We found Vote Loss slightly better on average and used it for all reported results.

3.6.4   LP Decomposition

Frequently our linear programs can be *decomposed* into smaller problems that can be solved independently. If there exists a subset of linear constraints $L' \subset L$ that contain a set of variables $V' \subset V$ such that $\forall_{v \in V', l \in L/L'} \ v \notin l$, then L' together with the terms in the cost function containing the variables V' can be solved as a separate LP.

We can also reduce running time by observing that, for any such "sub-LP", it is easy to set each variable c_v to ω_c/ω_{M_c} (yielding the minimum possible cost of 0) and check if the constraints are satisfied—if they are, the optimal solution is found without invoking the linear solver. Together, these techniques allowed us to solve most LPs one or two orders of magnitude faster in our experiments (almost always within a matter of seconds), taking more than a minute to solve on a modest desktop machine only when the presence of tens of thousands of constraints prevent meaningful decomposition.

3.6.5   Tie Breaking

We must also address "ties" between claims with the same number of votes. If the linear solver is allowed to break these arbitrarily, the results may vary from solver to solver. This is of particular concern when using a chain of solvers (our experiments used Microsoft Solver Foundation (MSF) simplex $\rightarrow$ lp_solve simplex $\rightarrow$ MSF interior point) to enable "fallback" when one solver fails and consistent behavior is required. To handle this we identify pairs of claims with the same number of votes in each decomposed LP, multiplying the votes of one by $1 + 10^{-10}$ and repeating until no pair of claims is tied. Which claim gets slightly boosted depends upon a "precedence" that is assigned randomly at the start of the experiment.

3.6.6   "Unknown" Augmentation

Augmenting our data with "Unknown" claims ensures that every LP is feasible and can be used to model our ignorance given a lack of sufficient information or conflicting constraints. An Unknown claim U_M is added to every mutual exclusion set M (but invisible to the fact-finder) and represents our belief that *none* of the claims in M is sufficiently supported. Now we can write the mutual exclusion constraint for M as $U_M + \sum_{c \in M} c_v = 1$. When propositionalizing first-order logic (FOL), if a disjunctive clause contains a nonnegated literal for a claim c, then we add $\vee\, U_{M_c}$ to the clause. For example, $Age(John, 35) \Rightarrow Age(Tom, 40)$ becomes $Age(John, 35) \Rightarrow Age(Tom, 40) \vee Age(Tom, Unknown)$. The only exception is when the clause contains claims from only one mutual exclusion set (e.g., "I know Sam is 50 or 60"), and so the LP can be infeasible only if the user directly asserts a contradiction (e.g., "Sam is 50 *and* Sam is 60"). The Unknown itself has a fixed number of votes that cannot be lost; this effectively "smooths" our belief in the claims and imposes a floor for believability. If $Age(Kim, 30)$ has 5 votes, $Age(Kim, 35)$ has 3 votes, and $Age(Kim, Unknown)$ is fixed at 6 votes, we hold that Kim's age is unknown due to lack of evidence. The number of votes that should be given to each Unknown for this purpose depends, of course, on the particular fact-finder and ω function used; in our experiments, we are not concerned with establishing ignorance and thus assign 0 votes.

3.7 Experimental Results

To evaluate our extensions to fact-finding, both the generalization of the fact-finders themselves and the application of constraints to encode prior knowledge, our experiments applied a number of state-of-the-art fact-finding algorithms to both real-world and semi-synthetic datasets. We considered both extensions separately, finding each was independently able to improve the accuracy of trust decisions by incorporating different types of background knowledge into the fact-finding process, and then combined these orthogonal components into a joint model able to achieve significantly better results than were possible using either alone.

3.7.1 Data

A number of real-world datasets are used, including two (Population and Biography) extracted from Wikipedia infoboxes [287] (semi-structured tables with various fields within Wikipedia articles). An example of an infobox for the city of Laguna Beach is shown in Figure 3.4.

City of Laguna Beach	
— City —	
Country	United States
State	California
County	Orange
Area	
- Total	9.7 sq mi (25.2 km^2)
- Land	8.8 sq mi (22.9 km^2)
- Water	0.9 sq mi (2.3 km^2)
Population (2000)	
- Total	23,727
- Density	2,683.5/sq mi (1,036.1/km^2)

FIGURE 3.4: Example of a Wikipedia Infobox.

3.7.1.1 Population

Pasternack and Roth [202] collected Wikipedia infoboxes for settlements (Geobox, Infobox Settlement, Infobox City, etc.) to obtain 44,761 population claims qualified by year (e.g., triples such as (Denver, 598707, 2008)) from 171,171 sources ("editors", in Wikipedia parlance), with a test set of 308 "true" claims taken from U.S. census data (omitting the many cases where editors did not contest the population, or where all claims in Wikipedia were wrong). To allow for a direct comparison between generalized fact-finding and declarative prior knowledge, we use the population dataset across both sets of experiments and for the combined joint model as well.

3.7.1.2 Books

For generalized fact-finding, we also have Yin et al.'s [290] Books dataset, extracted from online bookstore Websites. The Books dataset is a collection of 14,287 claims of the authorship of various books by 894 Websites, where a Website asserts that a person was an author of a book (e.g., (Bronte, "Jane Eyre")) explicitly by including them in the list of authors, or implicitly asserts a person was *not* an author (e.g., (¬Bronte, "Jane Eyre")) by omitting them from the list (when at least one other Website lists that person as an author of the book—if nobody lists a person as an author, his nonauthorship is not disputed and can be ignored). The test set is 605 true claims collected by examining the books' covers.

3.7.1.3 Biography

For our declarative prior knowledge experiments, Pasternack and Roth [202] created the Biography dataset by scanning Wikipedia infoboxes to find 129,847 claimed birth dates, 34,201 death dates, 10,418 parent-child pairs and 9,792 spouses as reported by 1,167,494 editors. To get "true" birth and death dates, we extracted data from several online repositories (after satisfying ourselves that they were independent and not derived from Wikipedia!), eliminating any date these sources disagreed upon, and ultimately obtained a total of 2,685 dates to test against.

3.7.1.4 American vs. British Spelling

Finally, we examined a domain where the truth was plainly subjective and thus the user's prior knowledge is essential: identifying the "correct" spelling of words given 209,189 articles from the British National Corpus, *The Washington Post* and Reuters written by 9,387 distinct authors [202].

3.7.2 Experimental Setup

For our experiments we used a number of state-of-the-art fact-finding algorithms: Sums/Hubs and Authorities (**Sum**), 3-Estimates (**3Est**), simplified TruthFinder (**TF^s**), "full" TruthFinder (**TF^c**), Average·Log (**A·L**), Investment with $g = 1.2$ (**Inv$^{1.2}$**), and PooledInvestment with $g = 1.4$ (**Pool$^{1.4}$**). The voting baseline (**Vote**) simply chooses the claim asserted by the most sources. The number of iterations used for each fact-finder was fixed at 20. To evaluate accuracy, after the final iteration we looked at each mutual exclusion

TABLE 3.2: Experimental Results for Tuned Assertion Certainty.

Data	Weights	Vote	Sum	3Est	TF^c	A·L	Inv$^{1.2}$	Pool$^{1.4}$
Pop	Unweighted	81.49	81.82	81.49	84.42	80.84	87.99	80.19
Pop	Tuned	81.90	82.90	82.20	87.20	83.90	90.00	80.60
Pop	Best	81.82	83.44	82.47	87.66	86.04	90.26	81.49

set M and predicted the highest-belief claim $c \in M$ (other than u_M, if applicable), breaking ties randomly, and checked if it was the true claim t_M. We omitted any M that did not contain a true claim (all known claims are false) and any M that was trivially correct (containing only one claim [other than u_M, if applicable]).

3.7.3   Generalized Fact-Finding

3.7.3.1   Tuned Assertion Certainty

A user modifying a field of interest in an infobox (e.g., the *population_total* field) is clearly asserting the corresponding claim ("population = x"), but what if he edits another part of the infobox, or something else on the page? Did he also read and approve the fields containing the claims we are interested in, implicitly asserting them? We can simply consider only direct edits of a field containing a claim to be an assertion of that claim, but this ignores the large number of potential assertions that may be implicit in an editor's decision *not* to change the field.

This may be considered either uncertainty in information extraction (since we are not able to extract the author's true intent) or uncertainty on the part of the authors (an editor leaves a field unaltered because he believes it is "probably" true). In either case, we can weight the assertions to model this uncertainty in the generalized fact-finder. The information extractor provides a list of all edits and their types (editing the field of interest, another field in the infobox or elsewhere in the document), and each type of edit implies a different certainty (a user editing another field in the infobox is more likely to have read and approved the neighboring field of interest than a user editing a different portion of the document), although we do not know what those levels of certainty are. These can be discovered by tuning with a subset of the test set and evaluating on the remainder, varying the relative weights of the "infobox", "elsewhere" and "field of interest" assertions. The results are shown in Table 3.2. Note that all values are percent accuracy. In the "unweighted" case only direct edits to the "field of interest" are considered, and "infobox" and "elsewhere" edits are ignored (giving all edits equal weight fares much worse).

We tuned over 208 randomly-chosen examples and evaluated the remaining 100, repeating the experiment ten times. We also tuned (and tested) with all 308 labeled examples to get the "Best" results, only slightly better than

TABLE 3.3: Experimental Results for Uncertainty in Information Extraction.

Data	Assertions	Vote	Sum	3Est	TF^c	A·L	Inv$^{1.2}$	Pool$^{1.4}$
Pop	Unweighted	71.10	77.92	71.10	78.57	76.95	78.25	74.35
Pop	Generalized (Weighted)	**76.95**	**78.25**	**76.95**	**80.19**	**78.90**	**84.09**	**78.25**
Books	Unweighted	80.63	77.93	80.74	80.56	79.21	77.83	81.20
Books	Generalized (Weighted)	**81.88**	**81.13**	**81.88**	**82.90**	**81.96**	**80.50**	**81.93**

those from legitimate tuning. As expected, assigning a smaller weight to the "infobox" assertions (relative to the "field of interest") and a much lower weight to the "elsewhere" assertions yielded the greatest results, confirming our common-sense assumption that edits close to a field of interest confer more supervision and implicit approval than those elsewhere on the page. We find a significant gain across all fact-finders, notably improving the top Investment result to 90.00%, demonstrating that generalized fact-finders can dramatically increase performance.

3.7.3.2 Uncertainty in Information Extraction

We next consider the case where the information extractor is uncertain about the putative claims, but provides an (accurate) estimate of $\omega_u(s, c) = P(s \rightarrow c)$, the probability that source s made a given claim c.

For the Population dataset, we augment each mutual exclusion set M with an additional (incorrect) claim, ensuring $|M| \geq 2$. For each assertion $s \rightarrow c$ we select another $c' \in M_c$, and draw a p from a Beta(4,1) distribution ($\mathbb{E}(p) = 0.8 \Rightarrow 20\%$ chance of error). We then set $\omega_u(s, c) = p$ and $\omega_u(s, c') = 1 - p$. In the unweighted case (where edge weights must be 0 or 1), we keep the edge between s and c if $p \geq 0.5$, and replace that edge with one between s and c' if $p < 0.5$.

For the Books dataset, each mutual exclusion set had exactly two claims (a person is either an author of a book or he is not) and thus did not require augmentation. Here we drew p from a Beta(2,1) distribution ($\mathbb{E}(p) = 2/3$), corresponding to a greater (33%) chance of error. Our results are shown in Table 3.3; on both datasets, generalized fact-finders easily outperform their standard counterparts.

3.7.3.3 Groups as Weighted Assertions

Using the Population data we considered three groups of editors: administrators, blocked users and regular users with at least one template on their user page (intended to capture more serious editors). To keep things simple, we allowed each user to belong to at most one of these groups—if an adminis-

TABLE 3.4: Experimental Results for Groups Using Weighted Assertions.

β	Vote	Sum	3Est	TF^c	A·L	Inv$^{1.2}$	Pool$^{1.4}$
(No groups)	81.49	81.82	81.49	84.42	80.84	87.99	80.19
0.7	**84.09**	**84.09**	**84.42**	**85.71**	**84.74**	84.74	**83.44**
0.5	83.77	**84.09**	**84.42**	85.06	84.09	87.01	82.79
0.3	82.47	83.77	83.77	84.74	83.77	87.01	82.79
0.00001	83.44	82.14	83.44	84.42	81.49	**88.96**	80.51

trator had been blocked, he nonetheless belonged to the administrator group; if an otherwise "regular" user were blocked, he (of course) belonged to the blocked group. Given that administrators are promoted to that position by being trusted by other Wikipedia editors, and that blocked users are blocked by trusted administrators for (presumable) misbehavior, we expected that administrators would be relatively trustworthy on the whole, while blocked users would be more untrustworthy, with serious editors somewhere in between. We then encoded these groups as weighted assertions, using ω_g with arbitrarily chosen β parameters, as shown in Table 3.4. We see improved performance with all β values tested, with the exception of the Investment algorithm, which requires a much lower β; we can conclude from this that β should be tuned independently on each fact-finder for best results.

3.7.3.4 Groups as Additional Layers

We next took the same three groupings of editors (administrators, blocked users and regular users) and added them as a third layer in our generalized fact-finders, continuing to use the same Population dataset as previously. For most fact-finders, we can directly adapt the T and B functions as U and D functions, respectively, though this excludes PooledInvestment (which depends on mutual exclusion sets) and 3-Estimates (whose "claim difficulty" parameters are not readily extended to groups). In the former case, we can calculate the trustworthiness of the groups in the third layer as a weighted average of the trustworthiness of its members, giving us $U_3^i(g) = \sum_{s \in g} U_2^i(s)/|g|$, where g is a group and $|g|$ is the number of sources it contains. Likewise, we can calculate the trustworthiness a source inherits from its groups as the weighted average of the groups' trustworthiness, giving $D_2^i(s) = \sum_{g \in G_s} D_3^i(g)/|G_s|$, where G_s is the set of groups to which source s belongs (recall that, since there are three layers, $D_3^i(g) = U_3^i(g)$). We can use these new U_3 and D_2 functions to handle the interaction between the group layer and the source layer, while continuing to use an existing fact-finder to mediate the interaction between the source layer and claim layer. We apply this hybrid approach to two fact-finders, giving us Inv$^{1.2}$/Avg, and Pool$^{1.4}$/Avg. Finally, note that regardless of the choice of D_2, we are discarding the trustworthiness of each source as established by its claims in favor of the collective trustworthiness of its groups, an information bottleneck. When we have ample claims for a

TABLE 3.5: Experimental Results for Groups as an Additional Layer.

Description	Sum	TF^c	A·L	Inv$^{1.2}$	Inv$^{1.2}$/Avg	Pool$^{1.4}$/Avg
No Groups	81.82	**84.42**	80.84	87.99	87.99	80.19
Group Layer	83.77	83.44	**84.42**	83.44	88.64	64.94
Group Layer with D_2^{smooth}	**84.74**	84.09	82.79	**88.96**	89.61	**84.74**
Tuned + Group Layer	**86.10**	83.30	**87.00**	**88.50**	90.00	77.90
Tuned + Group Layer with D_2^{smooth}	83.20	**85.30**	84.20	87.40	90.00	**83.50**

source, its group membership is less important; however, when there are few claims, group membership becomes much more important due to the lack of other "evidence". The previously described D_j^{smooth} captures this idea by scaling the impact of groups on a source by the (weighted) number of claims made by that source. We show results both with and without this smoothing in Table 3.5.

Except for TruthFinder, group information always improves the results, although "smoothing" may be required. We also tuned the assertion certainty as we did in Table 3.2 in conjunction with the use of groups; here we find no relative improvement for Investment or TruthFinder, but gain over both tuning and groups alone for all other fact-finders.

3.7.4 Constrained Fact-Finding

3.7.4.1 IBT vs. L+I

We can enforce our prior knowledge against the beliefs produced by the fact-finder in each iteration, or we can apply these constraints just once, after running the fact-finder for 20 iterations without interference. By analogy to Punyakanok et al. [212], we refer to these approaches as inference based training (IBT) and learning + inference (L+I), respectively. Our results show that while L+I does better when prior knowledge is not entirely correct (e.g., "Growth" in the city population domain), generally performance is comparable when the effect of the constraints is mild, but IBT can outperform when prior knowledge is vital (as in the spelling domain) by allowing the fact-finder to learn from the provided corrections.

3.7.4.2 City Population

Our "common-sense" knowledge is that population grows over time ("Growth" in Table 3.6); therefore, $\forall_{v,w,x,y,z} pop(v,w,y) \wedge pop(v,x,z) \wedge y < z \Rightarrow x > w$. Of course, this often does not hold true—cities can shrink—but performance was nevertheless superior to no prior knowledge whatsoever. The

TABLE 3.6: Constrained Fact-Finding Results. $\emptyset$ Indicates No Prior Knowledge.

Dataset	Prior Knowledge	Vote	Sum	3Est	TF^s	TF^c	A·L	$Inv^{1.2}$	$Pool^{1.4}$
Pop	$\emptyset$	81.49	81.82	81.49	82.79	84.42	80.84	**87.99**	80.19
Pop	Growth$_{IBT}$	82.79	79.87	77.92	82.79	**86.36**	80.52	85.39	79.87
Pop	Growth$_{L+I}$	82.79	79.55	77.92	83.44	85.39	80.52	**89.29**	80.84
Pop	Larger$_{IBT}^{2500}$	85.39	85.06	80.52	86.04	87.34	84.74	**89.29**	84.09
Pop	Larger$_{L+I}^{2500}$	85.39	85.06	80.52	86.69	86.69	84.42	**89.94**	84.09
SynPop	$\emptyset$	73.45	87.76	84.87	56.12	87.07	**90.23**	89.41	90.00
SynPop	Pop±8%$_{IBT}$	88.31	95.46	92.16	**96.42**	95.46	96.15	95.46	**96.42**
SynPop	Pop±8%$_{L+I}$	88.31	94.77	92.43	82.39	95.32	95.59	**96.29**	96.01
Bio	$\emptyset$	89.80	89.53	89.80	73.04	**90.09**	89.24	88.34	90.01
Bio	CS$_{IBT}$	89.20	89.61	89.20	72.44	89.91	89.35	88.60	**90.20**
Bio	CS$_{L+I}$	89.20	89.61	89.20	57.10	90.09	89.35	88.49	**90.24**
Bio	CS+Decades$_{IBT}$	90.58	90.88	90.58	80.30	91.25	90.91	90.02	**91.32**
Bio	CS+Decades$_{L+I}$	90.58	90.91	90.58	69.27	90.95	90.91	90.09	**91.17**
Spell	$\emptyset$	13.54	9.37	11.96	**41.93**	7.93	10.23	9.36	9.65
Spell	Words$_{IBT}^{100}$	13.69	9.02	12.72	**44.28**	8.05	9.98	11.11	8.86
Spell	Words$_{L+I}^{100}$	13.69	8.86	12.08	**46.54**	8.05	9.98	9.34	7.89
Spell	CS+Words$_{IBT}^{100}$	35.10	31.88	35.10	56.52	29.79	32.85	73.59	**80.68**
Spell	CS+Words$_{L+I}^{100}$	35.10	31.72	34.62	**55.39**	22.06	32.21	30.92	29.95

L+I approach does appreciably better because it avoids forcing these sometimes incorrect constraints onto the claim beliefs while the fact-finder iterates (which would propagate the resulting mistakes), instead applying them only at the end where they can correct more errors than they create. The sparsity of the data plays a role—only a fraction of cities have population claims for multiple years—and those that do are typically larger cities where the correct claim is asserted by an overwhelming majority, greatly limiting the potential benefit of our Growth constraints. We also considered prior knowledge of the relative sizes of some cities, randomly selecting 2500 pairs of them (a, b), where a was more populous than b in year t, asserting $\forall_{x,y} pop(a, x, t) \wedge pop(b, y, t) \Rightarrow x > y$. This "Larger" prior knowledge proved more effective than our oft-mistaken Growth constraint, with modest improvement to the highest-performing Investment fact-finder, and Investment$_{L+I}$ reaches **90.91%** with 10,000 such pairs.

3.7.4.3   Synthetic City Population

As our real-world data was sparse, we created a synthetic dataset to determine how effective common-sense knowledge would be in the presence of "dense" data. We chose 100 random (real) cities and created 100 authors whose individual accuracy a was drawn uniformly from $[0, 1]$. Between 1 and 10 claims (also determined uniformly) were made about each city in each year from 2000 to 2008 by randomly selected authors. For each city with true population p and year, four incorrect claims were created with populations selected uniformly from $[0.5p, 1.5p]$, each author claiming p with probability a and otherwise asserting one of the four incorrect claims. Our common-sense knowledge was that population did not change by more than 8% per year (also tested on the Wikipedia dataset but with virtually no effect). As with "Growth", "Pop±8%" does not always hold, but a change of more than 8% is much rarer than a shrinking city. These constraints greatly improved results, although we note this would diminish if inaccurate claims had less variance around the true population.

3.7.4.4   Basic Biographies

Our common sense ("CS") knowledge was nobody dies before they are born, people are infertile before the age of 7, nobody lives past 125, all spouses have overlapping lifetimes, no child is born more than a year after a parent's (father's) death, nobody has more than two parents, and nobody is born or dies after 2008 (the "present day", the year of the Wikipedia dump). Applying this knowledge roughly halved convergence times, but had little effect on the results due to data sparsity similar to that seen in the population data—while we know many birthdays and some death dates, relatively few biographies had parent-child and spouse claims. To this we also added knowledge of the decade (but not the exact date) in which 15,145 people were born ("CS+Decades"). Although common sense alone does not notably improve results, it does very well in conjunction with specific knowledge.

3.7.4.5   American vs. British Spelling

Prior knowledge allows us to find a truth that conforms with the user's viewpoint, even if that viewpoint differs from the norm. After obtaining a list of words with spellings that differed between American and British English (e.g., "color" vs. "colour"), we examined the British National Corpus as well as *Washington Post* and Reuters news articles, taking the source's (the article author's) use of a disputed word as a claim that his spelling was correct. Our goal was to find the "true" British spellings that conformed to a British viewpoint, but American spellings predominate by far. Consequently, without prior knowledge the fact-finders do very poorly against our test set of 694 British words, predicting American spelling instead in accordance with the great majority of authors (note that accuracy from an American perspective is 1−"British" accuracy). Next we assumed that the user already knew the correct spelling of 100 random words, removing these from the test set, of course, but with little effect. Finally, we added our CS knowledge: if a spelling a is correct and of length ≥ 4, then if a is a substring of b, $a \Leftrightarrow b$ (e.g., colour $\Leftrightarrow$ colourful). Furthermore, while we do not know a priori whether a spelling is American or British, we do know if e and f are different spellings of the same word, and, if two such spellings have a chain of implication between them, we can break all links in this chain (while some American spellings will still be linked to British spellings, this removes most such errors). Inter-

TABLE 3.7: Experimental Results for the Joint Framework.

Prior Knowledge	Group Layer	Sum	TF^c	A·L	Inv$^{1.2}$	Inv$^{1.2}$/Avg	Pool$^{1.4}$/Avg
$\emptyset$	No Groups	81.82	84.42	80.84	87.99	87.99	80.19
$\emptyset$	Unsmoothed	83.77	83.44	84.42	83.44	88.64	64.94
$\emptyset$	D_2^{smooth}	84.74	84.09	82.79	88.96	89.61	84.74
Larger$_{IBT}^{2500}$	No Groups	85.06	87.34	84.74	89.29	89.29	84.09
Larger$_{L+I}^{2500}$	No Groups	85.06	86.69	84.42	89.94	89.94	84.09
Larger$_{IBT}^{2500}$	Unsmoothed	87.34	86.04	**86.69**	85.71	89.94	72.40
Larger$_{IBT}^{2500}$	D_2^{smooth}	**87.99**	**87.66**	**86.69**	**90.58**	**90.26**	**87.99**

estingly, common sense alone actually *hurts* results (e.g., PooledInvestment (IBT) gets 6.2%), as it essentially makes the fact-finders more adept at finding the predominant American spellings! However, when some correct spellings are known, results improve greatly and demonstrate IBT's ability to spread strong prior knowledge, easily surpassing L+I. Results improve further with more known spellings (PooledInvestment gets **84.86%** with CS+Words$_{IBT}^{200}$).

3.7.5   The Joint Generalized Constrained Fact-Finding Framework

Our final experiments combine generalized and constrained fact-finding to create the full joint framework, capable of leveraging a very broad set of background and domain knowledge in our trust decision. We again use the Population dataset, applying the Larger^{2500} declarative prior knowledge set to generalized fact-finders using the Wikipedia editor group information (administrator, normal user, blocked user) encoded as an additional layer. The results in Table 3.7 show a significant and consistent gain using the joint framework with D_2^{smooth} across all fact-finders (with IBT; L+I results [not shown] were only slightly lower). The top result from the Investment fact-finder rises to 90.58%, up from 89.61% using only group information, or 89.94% using only declarative prior knowledge, while even the very simple Sums fact-finder achieves a respectable 87.99% performance, up from 81.82% with no background knowledge of any kind.

3.8   Conclusion

Generalized Constrained Fact-Finding offers a framework for incorporating a broad range of knowledge into our trust decisions by augmenting fact-finding algorithms, both by generalizing the fact-finders themselves and by constraining them with declarative prior knowledge. Generalized fact-finding allows us to encode factors such as information extraction and source uncertainty, similarity between the claims, and source groupings and attributes, with substantial and consistent performance gains across a broad range of fact-finders. Simultaneously, declarative prior knowledge, expressed as constraints over the belief in the claims, proves vital when the user's subjective truth differs from the norm, as it did in the Spelling domain, and even in other experiments where the "truth" is less contested both common-sense and specific knowledge provided significant benefit; moreover, as the constraints are enforced by a linear program, the framework remains polynomial-time, an essential characteristic when dealing with real-world "Web-scale" data. As both generalized and constrained fact-finding are orthogonal, they may be readily used together, achieving better results than are possible with either method alone and allowing the full breadth of our available information to be jointly leveraged in determining the oft-subjective truth in the presence of a morass of conflicting information.

Chapter 4

Web Credibility Assessment

4.1   Introduction

Millions of users are already using the Internet as the key source of information in their everyday lives. Personal finance, education or security are just three examples of domains in which the use of the Internet becomes almost obvious, and their number is rapidly growing. Medicine might be a glaring example of a kind of content where the users, not relying on credible sources, are virtually putting their health in jeopardy. The more the users rely on the information found on the Web, the more important it is to provide them with tools for efficient credibility assessment and it is more costly to be exposed to information that is not credible. Not every user possesses enough experience or knowledge to make correct assessments.

In this chapter we are introducing the idea of credibility on the Web, focusing specifically on the credibility of the Websites. We present phenomena to be encountered and difficulties to be dealt with while researching Web credibility prior to presenting experience in building an automatic credibility classifier. We start with introducing the credibility and related terms definitions. Afterwards, the process of data collection for research purposes is covered together with practical hints concerning this subject. The section devoted to the data analysis covers the most likely problems, e.g., rater cognition bias, controversy or best methods of crowdsourced ratings aggregation. The chapter concludes with a section devoted to automatic credibility classification based on Website features.

4.2   Web Credibility Overview

4.2.1   What Is Web Credibility?

Credibility is a very complex phenomenon. First of all, it is subjective in its nature and, therefore, prone to bias and social influence. Moreover, credibility can be analyzed as a multidimensional phenomenon. Not only is credibility itself composed of various elements such as trustworthiness and expertise, there can also be different types of credibility: presumed, reputed, surface, experienced [67]. Credibility can be simply defined as believability [72] or — in more sophisticated way — as a property of information that makes the receivers believe that it is true. In the Reconcile project the concept of credibility has been introduced in order to avoid the concept of truth as there is no philosophical agreement on its definition.

Though credibility is subjective and – as the Fogg's Prominence-Interpretation theory says - can be perceived and interpreted differently by various receivers, it also shows some objective aspects, as the consensus on credibility is possible to be reached among receivers (according to Reconcile team research, it happens in at least 30% of cases).

Evaluation of web credibility requires achieving the agreement between several Web users about the level of the content credibility. This solution supports effectively the property measure as it lowers its sensitivity to manipulation done by an individual.

Moreover, credibility is sometimes described as a perceived quality, thus when discussing particular Web credibility, one is always discussing the perception of the Web credibility, not the credibility itself [72].

This perception results from simultaneous evaluations of multiple dimensions (see, e.g., [25]). The two major components are trustworthiness of the source (defined by the terms *well-intentioned, truthful, unbiased*) and expertise of the receiver (defined by terms such as *knowledgeable, experienced, competent*) [72].

The broad definition of *credibility* implies that particular Webpages can be classified as "credible" from various reasons: one can be visually attractive and have a high responsiveness, while another one might be modest, but with long history and good reputation. Fogg proposes the following classification of different types of Web credibility:

- presumed credibility

- surface credibility

- earned (experienced) credibility

- reputed credibility [72, 138]

Presumed credibility is based on general assumptions and stereotypes held by an individual; it applies to various factors, such as domain identifiers (.gov, .edu, .org) or the amount of competitor information available on the site.

Surface credibility refers to a user's first impression and is influenced by such factors as professional (or nonprofessional) visual design or the amount of advertising located on the site. Fogg compares this type of credibility to "judging a book by its cover".

Earned (or experienced) credibility is built due to long-term relationships with the user and his or her evaluation of easiness of interacting with the site, responsiveness of the site, etc.

Reputed credibility is the least influenced by a user's personal judgement. It is created by seals of approval, links from credible sources (other sites or friends) and awards that the site may have won [72, 138].

In the earlier research terms *trust* and *credibility* were sometimes used as synonyms. Fogg suggests how to determine to which construct a particular phrase refers: "A number of studies use phrases such as trust in the information and trust in the advice. We propose that these phrases are essentially synonyms for credibility; they refer to the same psychological construct" [72].

4.2.2   Introduction to Research on Credibility

The empirical research on credibility dates back to the 1930s (see [208, 249] for review), but the concept is much older as it originated from Ancient Greece. According to Plato, credibility comes from knowing the truth, while Aristotle's opinion was that credibility is the communicator's talent to stimulate confidence and belief in what is being said. As Stacks and Salwen [249] conclude, this debate brought scholars to three sources of credibility:

- the source is credible when the audience perceives it as true

- the source is credible when it is communicated in a persuasive way, well-matched to particular audience

- the source is credible when the character of the audience makes it credulous.

More systematic and empirical (in the modern sense) research began in the first half of 20th century. Researchers were focused on the power of propaganda [132] and the need to determine whether radio advertising was more advantageous than the advertising in newspapers. The goal was to verify if the medium was perceived as a source itself and, therefore, was trusted or not. The key finding was that the reliance on a medium was proven to be the major credibility predictor (see, for example, [24, 110, 111, 224, 282]). The further studies on persuasiveness were conducted during World War II and after that by psychologist Carl Hovland and associates [100]. They examined how to persuade soldiers through wartime messages. The conclusion was that messages from more credible sources change attitudes more often than messages coming from sources with lower credibility.

In the late 1980s and early 1990s the approach to credibility was modified: the researchers, such as Gunther [88,89] and, Stamm and Dube [251], pointed out that the credibility is relational, i.e., it depends not only on the source of information, but mainly on the receiver's background. Gunter concluded four observations underlying this approach:

- Media credibility is a receiver assessment, not a source characteristic.

- Audience demographics, proposed as predictors of trust in media, have little theoretical basis and little empirical support.

- Situational factors often outweigh a more general skeptical disposition as predictors of credibility judgments.

- Group involvement will stimulate biased processing, affecting evaluations of messages and sources [249].

The continuation of this approach is a concept of "endorsed credibility" [66] — where credibility is based on "group and social engagement" and can be "poured" onto online sources. According to Flanagin and Metzger [66], "endorsed credibility in the digital media environment compensates for the relative anonymity of tools like the Web.... The means of sharing these assessments can assume many forms resulting in several variants of credibility, most notably conferred, tabulated, reputed, and emergent credibility".

This concept is also related to more technological approaches to establishing credibility, and trust metrics, within computer networks developed by information systems theorists (e.g., [304]). These propagation models "compute quantitative estimates of how much trust an agent a should accord to its peer b, taking into account trust ratings from other persons on the network".

This subchapter contains only a short summary of the past studies on credibility. For more robust information please refer to publications [177] and [71].

4.2.3   Current Research

Throughout the 20th and the beginning of the 21st century the importance of assessing credibility of the content increased rapidly, especially in last two decades, when, with the growing popularity of the Internet, the costs of information dissemination have lowered and, at the same time the accessibility to information has increased like never before. Problems have arisen because many sites operate without much oversight or editorial review. Information provided is not filtered by any professional gatekeepers (as with most printed publications) and there are no traditional indicators that would ensure the information credibility [177].

This is why research aimed at describing online content credibility evaluations has become a dynamically developing area. The bulk of most recent studies suggest that nowadays great emphasis is placed on credibility assessment in the context of main trends in information search. Many studies are dedicated to Wikipedia, one of the most popular and easily accessible information sources. Its specific design (with the possibility to add or edit content) makes it even a more interesting research subject because some information published there should not be trusted.

Concentrating on Internet users' characteristics, researchers have recently showed that perceived credibility of Wikipedia content is positively related to heuristic processing. Also familiarity with the topic of evaluated Wikipedia article was found to be moderating credibility evaluations [156, 222].

Another online service that can serve as a means of spreading information is Twitter — a popular social network and microblogging platform. Researchers measuring trust in different tweets (small posts) recently concluded that people base their credibility evaluations of information on visual design of the tweet author's profile [189]. Such findings may have many interesting practical applications. There have already been some attempts to create automated methods for assessing credibility of tweets based on tweet content characteristics and Twitter profile features [34, 91].

A broader area of research is dedicated to the process of determining behavior patterns of online information searches (in general). It focuses on preferred sources of information, and methods of assessing and verifying online information credibility [17, 181].

Consumer behavior studies, due to their different evaluation context, may be treated as another online credibility research branch. Most studies from this category focus on either Websites design or the impact of recommendations (recommender systems, reviews, etc.) on credibility evaluations [1].

More generally, research on Websites' credibility evaluation has so far focused on two basic areas: technical features of Websites and cognitive characteristics of Internet users. The former have been analyzed mainly in the context of Website thematic areas, usability and design. It is worth mentioning that design can be intuitively understood as an artistic manner but in the context of Websites' credibility evaluation it most often refers to a set of specific components included (or not) by the Website designer. It has been argued that people base their decisions concerning a Website's credibility on its graphical design rather than its information content [67].

However, in this chapter we will not focus on technical issues connected with Website design unless its perception may be moderated by Internet users' personal characteristics and, in such a sense, influence credibility evaluations.

4.2.4 Definitions Used in This Chapter

4.2.4.1 Information Credibility

Credibility is defined as a "property of information that makes the receivers believe that it is true". It cannot be denied that such properties exist; we hypothesize that they can be measured and used to classify information as credible or not credible (or indeterminate). Note that this definition of credibility avoids using the concept of truth; it relates rather to the mental state of receivers who "believe that received information is true". We also realize that credibility is subjective, as the various properties of information may be perceived and interpreted differently by various receivers, according to Fogg's Prominence-Interpretation theory. However, we also realize that credibility has objective aspects, since it is possible (as shown in our research, in at least 30% of cases) for receivers to reach a consensus about credibility.

One of the issues that remains to be clarified when we consider the above definition of credibility is who the receivers of the information are. This issue is significant in order to decide what it means to effectively support credibility evaluation, as the possible receivers can be intended beneficiaries of credibility evaluation support:

1. Individuals browsing the Web

2. Every Web user

3. A majority in a social group (the social group could be every Web user)

4. A topical knowledge community

5. Several social groups, who differ by fundamental social characteristics, such as race, age, gender, social class, wealth or nationality

The support of credibility evaluation that would be tailored just to an individual Web user is questionable. This approach would be liable to manipulation, as other individuals would use social influence to make others accept their versions of Truth. Credibility evaluation support tools cannot counteract such a process; rather they would tend to reinforce it, by reinforcing the beliefs of every individual user. Every other approach to credibility evaluation support requires achieving the agreement of several Web users regarding the evaluation of Web content credibility. What remains is to decide who should agree. Is agreement always possible?

4.2.4.2 Information Controversy

The possibility of disagreement is obvious for information that is controversial. The existence of controversy in science or in other forms of social discourse is not, in itself, a bad thing from the point of view of supporting credibility. An important function of credibility evaluation support is the distinction between a piece of information that can obtain an agreed-upon credibility evaluation from the controversial one. Information is controversial if users cannot agree on its credibility. The definition requires a definition of agreement, but at the moment we can consider a common-sense interpretation of near consensus (almost everyone agrees) as a right one. If we can diagnose that information is controversial, we will be still able to support a debate about its credibility. Such debate may not bring a resolution, but is useful in itself, as it can lead to the creation of new knowledge (as in science) or to the creation of social agreement (as in politics).

4.2.4.3   Credibility Support for Various Types of Information

The definition of credibility evaluation support requires the introduction of a concept of truth. In short, it is intuitive to say that supporting credibility evaluation means bringing this evaluation closer to the truth about the evaluated information. However, there are important philosophical questions concerning the concept of truth. Without going into details, let us state that if truth is indeed just the outcome of a social discourse (as postulated by post-structuralism), then information processing tools that aim to gather receivers' credibility evaluations, and to facilitate the discussion of credibility, should support the creation of post-structuralist truth, while remaining aware of the impact of context on credibility evaluations. However, the same applies to a scientific truth, which is a concept useful for information that can be verified based on reality. Credibility evaluation support for such information should aim at agreement with scientific truth, even if it needs to apply different methods to achieve this objective.

It is worth mentioning here a theory of truth that can also be applied to better understanding of credibility: Tarski's Semantic Theory of Truth[1]. In brief, this theory states that truth is a logical property of sentences, but it is relative to a model that expresses the semantics of these sentences. This view explains several apparent paradoxes of scientific truth, such as the disagreement between Newton's laws of motion and Einstein's theory of relativity. We can say that Newton's laws are valid if the model that expresses their semantics is limited to macroscopic objects in everyday conditions, but they are not valid for a model that takes into account microscopic objects and objects moving at very high speeds. Semantic Truth Theory lets us also think about credibility evaluation in a different way. There may be several models (or even sets of models) that are differently and independently used by people to evaluate truth: models for human beliefs, models for everyday facts, and models for specialized knowledge. By human beliefs we mean information that cannot be verified based on reality. By specialized knowledge we mean information that can be verified only by an expert or with the use of specialized equipment. By everyday facts we mean information that can be verified using ordinary intelligence and sense. While these distinctions may sometimes be fuzzy, they

---

[1]http://en.wikipedia.org/wiki/Semantic_theory_of_truth

TABLE 4.1: Semantic Models of Truth and Methods of Credibility Evaluation Support

Semantic truth Model	Methods of credibility evaluation support	Receivers who can benefit from Credibility evaluation support
Human beliefs	Discussion, with the goal of presenting diverse views	Several social groups, who differ by fundamental social characteristics, such as race, age, gender, social class or wealth, nationality
Everyday facts	Wisdom of crowds	Everyone
Specialized knowledge	Seeking evaluations by trusted, objective experts	Topical knowledge communities

are useful because they allow us to consider various methods of supporting credibility evaluation for noncontroversial information. In Table 4.1, the three models are shown along with proposed methods of credibility evaluation support and the receivers who could most benefit from these methods. Table 4.1 may not be comprehensive, but it serves to show that various approaches of credibility evaluation support can be attempted for various truth models and various intended audiences.

4.3 Data Collection

4.3.1 Collection Means

We propose three practical ways of obtaining a research credibility–related dataset. First, the most convenient, is reusing of an already available dataset. Second, make use of data stored by credibility evaluation supporting tools; and finally, collect a dataset by manually labeling the Websites, e.g., via crowdsourcing.

4.3.1.1 Existing Datasets

Despite the fact that reusing is cheaper and less time-consuming, not putting an effort into building one's own dataset have several shortcomings. A nontrivial issue one may encounter is that the available dataset does not fit one's needs, which can be severe as the list of publicly accessible credibility datasets is currently limited. Reusing a dataset also includes risk of having no access to the same versions of the content that was evaluated in the given dataset. As the Webpages are updated or abandoned, the possibility that there will be no way to reevaluate them is high.

One of publicly available datasets is one from a study on Web credibility and augmenting search results by Schwarz and Morris [237] and kindly shared by Microsoft[2]. In this dataset the examined 1000 Webpages were labeled a single time. Another, much bigger dataset, consisting of 6000 rated Webpages is available at the Webpage of a research project Reconcile[3]. This dataset is presented in Section 4.3.3.

4.3.1.2   Data from Tools Supporting Credibility Evaluation

Another approach to obtain a dataset consists of using external sources that provide credibility-related data although not in a structured form. By repeating the steps taken in the work of Sondhi et al. [247], one can look for lists of a priori credible or not credible content. Sondhi et al. used a list of credible medicine sites, accredited by Health on the Net foundation[4]. As proposed in Fritch [73] directories of prescreened pages by raters of a certain level of expertise can be referred to as sources of expert labeled data. An example of a directory of trusted pages assessed and chosen by librarians can be the IPL2 service pages[5]. On the contrary, a potential source of a priori non credible sites are lists of pages infected with malware or spyware. Such black lists are publicly available, e.g., the StopBadware organization[6]. Reputation systems focused on trust can also be used as a source for a credibility dataset, e.g., WOT (Web of trust), Reconcile, and Factlink. A wide overview of such tools is presented in Section 4.3.2.

4.3.1.3   Data from Labelers

Provided that one has available time and resources, it is a good idea to prepare a unique dataset tailored for the task's requirements. A set of Webpages chosen for study needs to be labeled according to its credibility, and this task needs to be performed with expertise and without a bias. A natural way of achieving such a goal is hiring domain experts to determine the credibility of the content of our interest. The number of experts is limited by our resources and the type of incentives used for the raters.

Another option is a judgment crowdsourcing [42], specifically carried out using crowdsourcing marketplaces, e.g., Amazon Mechanical Turk or Click-worker. Howe [101] introduced the definition of crowdsourcing as "the act of a company or institution taking a function once performed by employees and outsourcing it to an undefined (and generally large) network of people in the

---

[2]http://research.microsoft.com/en-us/projects/credibility/
[3]http://www.reconcile.pl
[4]http://www.hon.ch
[5]www.ipl2.org
[6]https://www.stopbadware.org/

form of an open call". In this special case, incentives for the labelers are also the key factor determining the size of the dataset. Another aspect of crowd-sourcing credibility is the unknown expertise level of the hired workers. Thus the gathered feedback on credibility of the given pages from single raters needs to be aggregated into final assessment labels. This is not a trivial task to perform, due to the unknown expertise of the workers as well as their motivation and honesty. The noise and spam in the crowdsourced data is a ubiquitous issue recently discussed in the field's literature [102, 242]. The manner of filtering the data and dealing with such labeler-related issues is discussed in the later sections of this chapter.

The risk of dishonest or biased workers applies as well to a crowdsourcing with hired workers as reputation systems. Both methods of obtaining data need to address a matter of credibility of the raters themselves [197]. Fabricated ratings, especially if the system enables giving feedback anonymously, can jeopardize the goal of gathering reliable data [138]. Another issue with data gathered via the crowdsourcing marketplace might be the sample of raters participating in the task. If one's study or work requires the raters' sample to be balanced, this issue needs to be addressed. Amazon Mechanical Turk user demographics in 2009 were an approximation of the U.S. Internet users population[7,8]. However, according to one of the most recognizable practitioners, Panos Ipeirotis, the profile of the Mechanical Turk users (workers) is changing, and the proportion of users coming from the United States dropped from almost 80% to about 50% in 2012 [102]. It should be expected that the distribution raters of demographic features will not be balanced.

4.3.2 Supporting Web Credibility Evaluation

There are many tools to help evaluate Website credibility. They cover one or both of two main issues with Web credibility as distinguished in Lazar et al. [138] and Schwarz and Morris [237]: credibility of information and Website security. First, users can be guided by some hints as to how to assess a Website by themselves. This way, using the same guidelines, experts can reconstruct their credibility evaluation. However, as is mentioned in Metzger [178], not many users will make the effort to perform such complex evaluation and the author suggests that credibility assessments should be for users, but not up to users. This can be done by sharing users experiences and opinions about Websites or checking reviews made by experts and institutions. There are also many databases and lists of automatically prescanned Websites. Another approach is to use seals or other signs of trust, security and credibility placed on Websites that can help with assessing. Some tools include using more than one method to evaluate site.

---

[7]http://www.behindtheenemylines.com/2009/03/turkerdemographicsvsinternet.html
[8]http://behindtheenemylines.blogspot.com/2010/03/newdemographicsofmechanicalturk.html

4.3.2.1   Support User's Expertise

Various tools are designed to help users in their own evaluation, providing them with the knowledge of what they should be aware of and what they should check. Such tools are, for example, checklists, question lists and high-lights, making users more conscious of credibility. A huge advantage of this tool is that users can learn such methods and naturally follow them later without any guidance. However, as mentioned earlier, this method requires (from a user) a lot of time and effort. These tools can be good for an expert's Web assessment, even if it is known that users rarely follow them.

Examples include the following:

- Widener University checklists[9]

- University of Maryland Library checklist[10]

- The Teaching Library, University of California, Berkeley, checklist[11]

- Health on the Net[12]

- Discern[13]

4.3.2.2   Crowdsourcing Systems

Also known as reputation, rating or recommendation systems, crowdsourcing systems use multiple users' reviews to provide assessment of Website credibility. Although users can receive recommended credibility evaluation from such tools without evaluating Website by themselves, they are encouraged to do it (e.g., by reputation, gamification or social network system).

Examples of these systems include the following:

- ReConcile[14]

- MyWot[15]

- Factlink[16]

- Hypothes.is[17]

---

[9]http://www.widener.edu/about/campus_resources/wolfgram_library/evaluate/original.aspx
[10]http://www.lib.umd.edu/binaries/content/assets/public/usereducation/evaluating-web-sites-checklist-form-fall-2012.pdf
[11]http://www.lib.berkeley.edu/TeachingLib/Guides/Internet/EvalForm.pdf
[12]http://www.hon.ch/
[13]http://www.discern.org.uk/
[14]http:// www.reconcile.pl
[15]https://www.mywot.com/
[16]https://factlink.com/
[17]http://hypothes.is/

4.3.2.3 Databases, Search Engines, Antiviruses and Lists of Pre-Scanned Sites

These solutions are based on premade lists of credible and/or not credible Websites, features or statements. Such lists are made by institutions, experts or other competent entities.

Such tools can, for example, inform about Websites from lists of malicious sites, a solution commonly used by antivirus and Internet protection software. More advanced tools can check Websites using particular features, content or links.

Examples include the following:

- Stop Badware[18]

- Norton ConnectSafe / Norton Safe Web / Norton Internet Security[19]

- AVG Link Scanner / AVG Internet Security / AVG AntiVirus[20]

- Avast! Internet Security / Avast! AntiVirus[21]

- VirusTotal[22]

- McAfee SiteAdvisor / various McAfee antivirus software[23]

4.3.2.4 Certification, Signatures and Seals

Some Websites are marked by authority and that can help in credibility assessment. If such a mark is verified, anyone can use it as a sign of credibility (if such authority is credible so whole assessment is transferred from a particular Website evaluation to a source/authority evaluation).

This can be done in many ways and at many levels. First, there are technical ways of providing information about Website security, for example, Security Socket Layer (SSL) certificates must be signed by a certificate authority and verified by browsers. Another example is the .gov domain Websites which are authorized by the General Services Administration. Second, Webpages can place a specific sign or seal of external authority. This way the Webpage can use "shared" trust from seal owner. Note that such seals are usually in a particular context, for example a site sealed about security can be secure and safe, but truthfulness of information presented is not guaranteed

Examples include:

- Norton Secured[24]

---

[18]https://www.stopbadware.org/
[19]https://safeweb.norton.com/
[20]http://www.avg.com/
[21]http://www.avast.com
[22]https://www.virustotal.com/
[23]http://www.siteadvisor.com
[24]http://www.symantec.com/page.jsp?id=seal-transition

- McAfee Secure[25]

- TRUSTe[26]

- BBB[27]

- Thawte[28]

- Trustwave[29]

- GeoTrust[30]

- Comodo[31]

4.3.3   Reconcile – A Case Study

Reconcile: Robust Online Credibility Evaluation of Web Content is a joint research project of two universities, Polish-Japanese Institute of Information Technology (PJIIT)[32] and École Polytechnique Fédérale de Lausanne (EPFL)[33], supported by the grant from Switzerland through the Swiss Contribution to the enlarged European Union. The Reconcile study is about how people evaluate the credibility of present Websites. The study gathers information not only about particular Websites results, but also about criteria of those evaluations together with knowledge about respondents' backgrounds. All this combined is designed to study how people judge different sites as credible or not. As a result of the project a dataset, Web Credibility Corpus (WCC), was gathered. This section describes the dataset and the process of collection. Subsequent parts of this chapter draw practical examples from the WCC dataset to present various phenomena likely to be encountered while researching credibility.

Goals were to build a midsize corpus of Webpages in English with credibility evaluations, in order to enable checking how accurately credibility can be predicted on the basis of extracted site features and to examine how context (i.e., topic, other pages within the given site, other sites on the same topic, task) influences a user's credibility evaluation. Other goals were to (1) enable the aggregation of credibility ratings for a Website based on evaluations of Webpages within it; derive credibility ratings for texts based on evaluations of statements by extracting the influence of text and other features related

---

[25]http://www.mcafee.com/us/mcafeesecure/about/legal/trustmark-info.html
[26]http://www.truste.com/products-and-services/enterprise-privacy/TRUSTed-websites
[27]http://austin.bbb.org/article/put-the-bbb-dynamic-seal-on-your-website-24428
[28]https://www.thawte.com/ssl/secured-seal/
[29]https://www.trustwave.com/trustedCommerce.php
[30]http://www.geotrust.com/ssl/ssl-site-seals/
[31]http://www.comodo.com/e-commerce/site-seals/secure-site.php
[32]www.pjwstk.edu.pl
[33]www.epfl.ch

to credibility, and identify heuristics for people to verify a Webpage credibility; (3) evaluate collaborative filtering-based approaches to recommend users credible Web content; (4) construct extensional definition of credibility, disentangle credibility and importance.

To recruit participants, the study used crowdsourcing marketplace Amazon Mechanical Turk[34]. To ensure best quality of the evaluations, several automatic validation mechanisms were implemented, e.g., minimum length of textual rating justification, minimum time of evaluation, provided links correctness. Additional manual evaluation was also performed and the final manual rejection rate amounted to 2%, that is, only 2% of tasks that passed automatic validation were eventually labeled as spam by hand and rejected. This, together with the relatively long time spent by the users on single evaluations (depicted in Figure 4.1) is taken as a sign of good data quality.

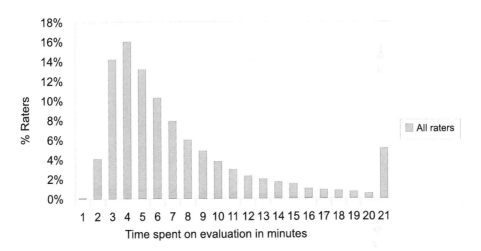

FIGURE 4.1: Distribution of respondent time spent on evaluation in minutes.

First, participants were asked about some basic information about themselves (demographics and psychology) and their Internet skills and experience (see Figure 4.1). After completing this questionnaire they proceed to the main part of experiment — site evaluation. They could evaluate up to 50 Websites. In the meantime, some of participants were asked to do an additional task — to answer a question related to the Website topic.

Websites were selected semimanually from three sources:

1. Google queries — pages selected from Google search results for several thematic queries

2. Web of trust categories — pages selected from WOT thematic categories

---

[34]www.mturk.com

3. Really Simple Syndication (RSS) feeds — pages selected from subscribed RSS feeds and followed for several weeks

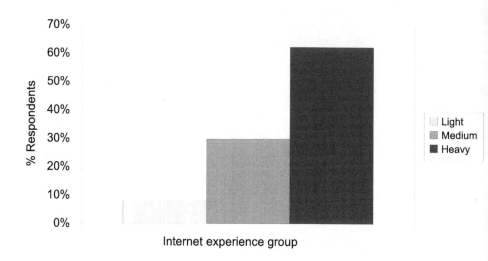

FIGURE 4.2: Web Credibility Corpus labelers by Internet experience level.

Participating in the study were 2405 different Amazon Mechanical Turk workers. The vast majority, i.e., 95%, came from United States, which was one of the conditions of the study. More than 60% of the workers were classified as heavy Internet users. The assignment to Internet experience groups was based on a questionnaire covering user Internet activity patterns and familiarity with Internet related terms, as is described in Kakol et al. [118]. The respondents were evaluating the credibility of subsets of a total of 5691 Webpages selected for the experiment with known and balanced distribution of credibility rating compared to the external system WOT. A total of 19,872 evaluations were collected. The average was 8 evaluations per worker and 8 evaluations per Webpage. Workers were evaluating the credibility on a 5-level scale:

1. Completely not credible

2. Mostly not credible

3. Somewhat credible, although with major doubt

4. Credible, with some doubt

5. Completely credible

Respondents' evaluation times varies from 57 seconds up to 1 hour and 25 minutes. More than 50% of evaluations were made in less than 6 minutes. Only 3 of 15,861 evaluations took less than 1 minute.

FIGURE 4.3: Credibility ratings distribution in Reconcile Web Credibility Corpus.

More than 40% of the respondents evaluated a presented site as 5 — completely credible and about 30% as 4 — credible with some doubt. Less than 10% of the evaluations were completely not credible (1) or mostly not credible (2). 15% of marks were 3 — somewhat credible, although with major doubt (see Figure 4.3).

The collected results are biased with distribution is skewed toward high credibility. More about biased credibility ratings can be found in Section 4.4.3.

All Websites were categorized according to their topic. Each topic had its own question to check participants' knowledge. Topics were grouped into 5 main categories: medicine, personal finance, healthy lifestyle, politics (with economy and ecology) and entertainment. Participants received random sites with information about their categories (see Figure 4.4).

Each participant evaluated the presented Website (selected at random from the database) providing three keywords to describe the Website, general credibility evaluation (using a 5-point scale), textual justification of their rating (open-ended question, participant must describe reasons for his evaluation) and links to related Websites that can help to evaluate visited Website. Additionally, evaluation of Website was done on four specific dimensions: Website appearance, author's expertise, author's intentions, information completeness (using a 5-point scale, see Figure 4.5). Respondents were also asked about their experience, strength of opinion and knowledge about the particular Website's topic.

As mentioned previously, along with general credibility, respondents evaluated other dimensions of credibility: presentation, knowledge, intentions and completeness. Results show that most of them are highly correlated (Spear-

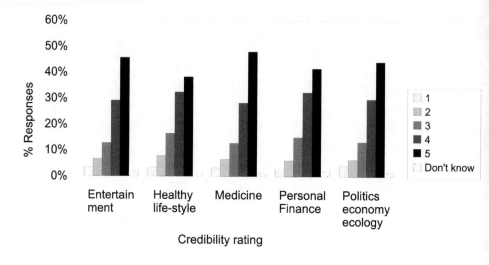

FIGURE 4.4: Credibility ratings distribution in different thematic categories in Reconcile Web Credibility Corpus.

TABLE 4.2: Spearman Correlation of Evaluated Dimensions in Reconcile Credibility Corpus

	Credibility	Presentation	Knowledge	Intentions	Completeness
Credibility	1	0.56	0.61	0.54	0.64
Presentation	0.56	1	0.53	0.37	0.53
Knowledge	0.61	0.53	1	0.49	0.63
Intentions	0.54	0.37	0.49	1	0.52
Completeness	0.64	0.53	0.63	0.52	1

man's rho above 0.5, see Table 4.2). Completeness is the most correlated dimension with credibility (0.64).

In the next section we present the analysis outcomes of the introduced Reconcile dataset.

4.4   Analysis of Content Credibility Evaluations

4.4.1   Subjectivity

Considering specifically the subjectivity of Web credibility and diverse demographics of the Internet users, the research into Web credibility becomes a complex task. Users' different outlooks, beliefs and needs make it complex. Good examples of this diversity are users requiring special attention, that is,

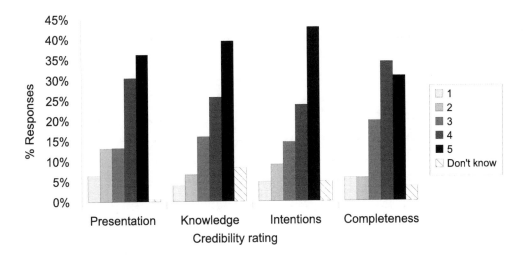

FIGURE 4.5: Distribution of other evaluated dimensions in Reconcile Web Credibility Corpus.

seniors or children might lack knowledge and experience needed for making a correct assessment [138]. Recent studies indicate also the differences in perception of credibility and trust even among different nations, as reported in work of Nielek et al., where data from a credibility reputation system (Web of Trust) together with a cross-national attitude survey, European Social Survey, were compared [195]. In other credibility related studies the differences in credibility assessment among participants of different socioeconomic backgrounds were also reported. Responses among different groups of gender, education and Internet experience have shown statistically significant differences [118]. Figures 4.6, 4.7 and 4.8 depict the observed differences, e.g., the respondents with higher Internet-related experience level tended to give lower credibility ratings in comparison to the low-Internet-experience respondents. The same tendency showed users with higher education level giving lower scores in comparison to lower education level user groups.

Subjective perception of credibility can lead us to the conclusion that the most suitable way to support credibility assessment consists in personalized recommendation of content according to the receiver's profile. Such attempts have already been made, e.g., in the work Seth, Zhang and Cohen, where a recommender system for credible blog posts was proposed. As reported in the work the results were promising and showed better accuracy in comparison with other recommendation methods [238]. However, it is not always the case that the Web credibility consists only of the subjective compound. Measuring credibility by gathering user ratings' just as in a reputation system, and amalgamating the responses into one answer seems applicable to a variety of cases. Such a case is the content that can be empirically examined against its

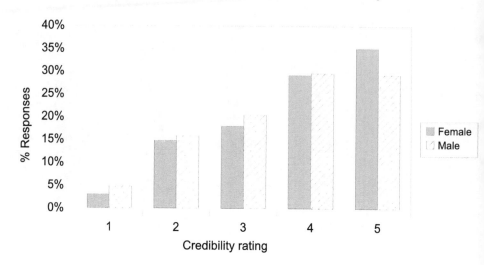

FIGURE 4.6: Distribution of credibility ratings by gender groups.

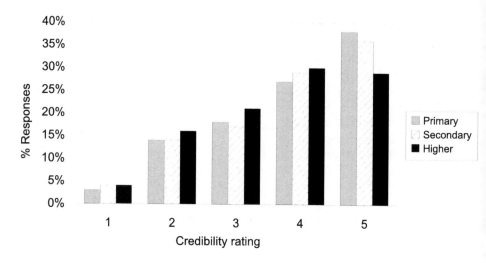

FIGURE 4.7: Distribution of credibility ratings by education level groups.

veracity. This can be concerning commonly agreed-upon scientific facts. However, controversy, manipulation or influential majorities imposing their points of view exist in science [53]. Another example can be human beliefs, which can surely have multiple subjective assessments of credibility.

Asking respondents to assess Web content and use their responses to generate average ratings, e.g., crowdsourcing or recommender systems, is acceptable and in the majority of cases the most convenient method. On the other hand,

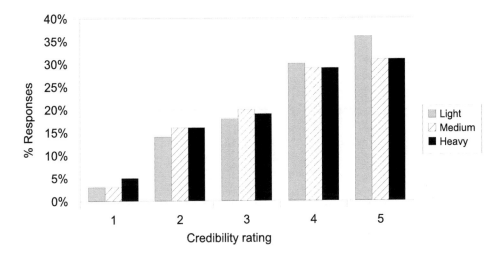

FIGURE 4.8: Distribution of credibility ratings by Internet experience user groups.

in the case of controversy of the content, the credibility assessment will remain unsettled.

4.4.2 Consensus and Controversy

An obvious way to assess credibility of a given Website is to ask an expert for an assessment. The content being assessed needs to be, of course, of the same domain as the field of expertise of the assessing expert. If a group of experts is assembled in order to make an assessment, their final decision should be by synergy smarter than the smartest of the experts in the group. However tempting the experts' assessments could be, the experts' unavailability can be a problem. Despite this fact, even a group of lay people of different expertise levels can still produce meaningful assessments according to the wisdom of crowds principle: "When our imperfect judgments are aggregated in the right way, our collective intelligence is often excellent" [254]. The concept of harvesting the wisdom of crowds became influential and widely used; however, it is still a subject of criticism, which covers the limitations of crowd wisdom and its proper application [131].

There are several requirements to be met in order to achieve a wise crowd. According to Surowiecki [254] these are diversity, independence and decentralization. In other words, the crowd needs to consist of individuals of different knowledge levels, who are expressing their private opinions with no influence from the others, by using their local knowledge and private information. These conditions may be difficult to satisfy due to frequent lack of robust control on the diversity of the pool of respondents in reputation systems. If the crowd does not meet all the conditions and specifically consists of uninformed members only, we cannot expect the most optimal decisions to be made. In terms of diversity and collaborative environments, there is also a possibility of situations in which the minority of well-informed experts is marginalized by the majority of less informed lay people, thus, the constructive feedback of the knowledgeable vanishes in the noise [80].

According to Lanier answers that the crowd is asked to give should be no more complicated than a single number or value [130]. Taleb draws attention to the limits of the crowd wisdom, which should not be applied to questions of complex outcomes and unknown distributions [256]. So what kind of task is it to assess web credibility? Surowiecki presents three types of problems the crowd is capable of solving: cognition, coordination and collaboration tasks. The credibility assessment is the cognition task in which the crowd is asked whether a given Website is credible. In this type of task well-formulated questions should have a single right answer [254], which unfortunately does not always apply to Web credibility assessment. In the majority of cases aggregation of the crowd responses works surprisingly well; however, there are some cases in which aggregation should be avoided, and we should accept the fact that we cannot produce meaningful assessment.

While evaluating the credibility, we need to consider the thematic category of the evaluated content as the crowd approach will work best with general knowledge content assessment. Considering the strong subjective component of credibility evaluation, several subjects need to be treated with caution. Such subjects are inherently controversial and cover sex, religion and other culture dependent taboos. Human beliefs are hardly subject to assessment that is expected to produce a single right answer. For example, it is yet impossible to objectively prove or disprove the existence of God. We can draw another example from Reconcile studies dataset. Among many categories of pages, which were labeled by crowdsourcing, there was "Cannabis", a category of pages concerning to the use of marijuana. The distribution of the credibility ratings gathered in this category visibly differs from the overall distribution and is depicted in Figure 4.9. The 'Cannabis' received almost uniform distribution of the ratings (except for rating 1). This is not surprising as the drug use related subjects are widely considered as "controversial".

Surowiecki says, "Diversity and independence are important because the best collective decisions are the product of disagreement and contest, not consensus or compromise" [254]. As mentioned in the above paragraphs, such a rule does not apply well to all kind of problems one would wish to solve using a wise crowd. In terms of Web content assessment one should carefully monitor the thematic categories of the items selected for evaluation. When the number of pages is high or the selection method is automated, this becomes a difficult task, due to the diversity of the content on the Web. At the aggregation step a personalized result reflecting the current user's standpoint and preference is presented. Another solution for such a dilemma is to monitor the level of agreement among the respondents for the assessed item. Figure 4.10 shows

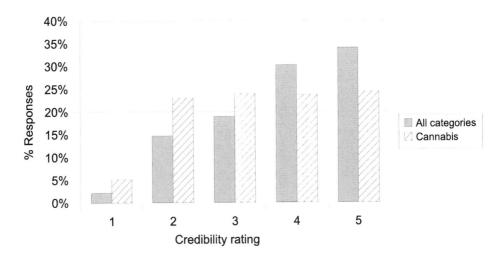

FIGURE 4.9: Distribution of credibility ratings in "Cannabis" category in comparison to overall distribution of credibility.

possible extreme ratings distribution. Perfect agreement on the left shows all the responses concentrated into one class. No agreement in the middle is depicted as a uniform distribution of the ratings and finally polarization is presented on the right. Perfect polarization depicted in Figure 4.10 shows even distribution of the ratings between two distant and opposite classes, what later in this section will be interpreted as a strong controversy.

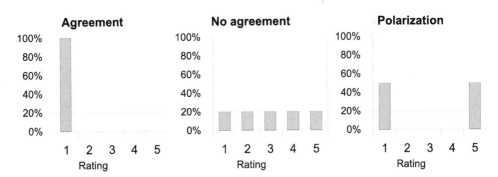

FIGURE 4.10: Levels of agreement depending on the ratings distribution.

Respondents rating Web content credibility are likely to be asked to place their perceptions on a scale defined by two polar opposites reflecting non-credibility and credibility. The concentration of those reported perceptions can be referred to as the agreement on the item's credibility [270]. Most likely the scale used to measure such opinion will be an ordered scale of the Likert-type on which measurement of the concentration or dispersion needs to be addressed in a proper way. Assuming an interval scale for the Likert categories and using, e.g., standard deviation poses a risk of reaching false conclusions [106]. Thus, consensus measures for ordinal variables should be used instead, in order to depict the extent of the inter-rater agreement. The agreement or ordinal consensus measures of Van der Eijk [270], Leik [140] and Tastle and Wierman [263] can be used for this purpose. Such agreement measures are typically normalized from 0, representing polarization, to 1, representing perfect agreement, thus effectively indicating the controversy.

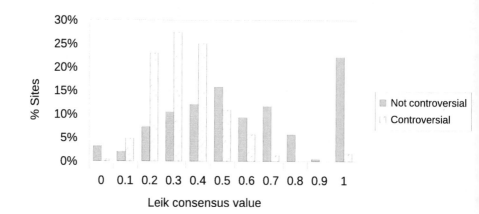

FIGURE 4.11: Distribution of Leik consensus values.

Given a training dataset covering controversy and credibility or trustworthiness assessments, it is feasible to build a controversy classifier. One possible source of training data is Wikipedia's Article Feedback Tool (AFT), which is an internal Wikipedia survey for article feedback, to engage readers in the assessment of article quality[35]. However, as depicted in Figure 4.11 consensus measures, specifically Leik consensus, perform well at discriminating controversial and not controversial pages based on the user ratings distributions. There is a visible concentration of controversial pages for consensus below 0.4 value, which is the the threshold for polarization.

An agreement measure can be used to monitor the potential polarization of the credibility perceptions of the raters. If strong polarization occurs, it might be a sign of controversy, a state in which two opposite outlooks on the same matter exist. Ratings concerning controversial content should not be aggregated. In terms of the Reconcile Web Credibility Corpus, the pages

---

[35]http://en.wikipedia.org/wiki/Wikipedia:Article_Feedback_Tool

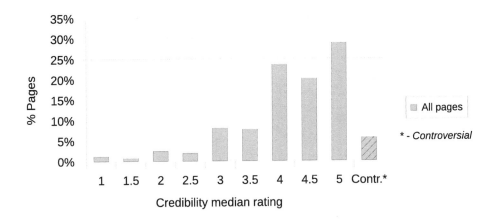

FIGURE 4.12: Reconcile Web Credibility Corpus median credibility ratings distribution, including "controversial" tag.

considered as controversial, based on the ratings distribution, amount to about 5% of all evaluated pages, as depicted in Figure 4.12.

The existence of controversy, that is, the existence of opposite credibility perceptions, is possible in the light of bounded rationality or Prominence-Interpretation (P-I) theory [67]. The controversies easily explained by P-I theory are manageable, and the wisdom of crowds approach can still be applied and ratings themselves aggregated. To the contrary to P-I explainable controversies, inherently controversial subjects, taboos and beliefs should be restrained from aggregation of the crowd ratings.

4.4.3 Cognitive Bias

4.4.3.1 Omnipresent Negative Skew – Shift Towards Positive

As mentioned in Section 4.2.4, there are several methods of credibility evaluation support depending on the adopted semantic truth approach. The one that we decided to focus on in our research is a wisdom of the crowds method of credibility evaluation deriving from the Everyday Facts semantic truth model. The very idea standing behind this approach is to gather as many credibility evaluations from diversified Web users as possible. This method, as with probably every other, has some limitations. It has been noted that the wisdom of crowds is biased. The typical finding is that people's evaluations tend to be overly positive, which leads to a negatively skewed distribution. It implies that most items are deemed credible, trustworthy, useful, interesting, etc.

This phenomenon has been observed in several cases. The most prominent example is the online auction platforms such as eBay, where users can leave feedback about their partner, after the transaction is completed. Most of these opinions (around 99%) are labeled as positive. This led many researchers to infer that both neutral comments and lacking comments are in fact implicit negative opinions [51, 51]. The online auctions constitute, however, a very particular example, since the bias in distribution of opinions in this case can be explained in strategic terms [19].

This explanation does not apply that well to other systems where users' opinions are being gathered. Kostakos [125] shows, for instance, that opinions about books on Amazon and BookCrossings are overly positive. Most items receive 5 out of 5 stars and most users vote 5 stars. On Amazon this effect is stronger in spite of elaborate control mechanisms. Kostakos stipulates that these very mechanisms compel users to avoid negative reviews. This, however, is not the whole story. The most straightforward cause can be a sample selection bias. Namely, people read and review only books that are potentially interesting to them. If they succeed in identifying the right content, there is no reason for them to give low ratings.

The analogous reasoning does not apply to the Reconcile study and credibility evaluations. Kakol and coauthors [118] use a balanced data set of Webpages and a balanced sample of Internet users who are asked to assess trustworthiness of the Webpages. The assignment of webpages to the subjects is random. Therefore, the sample selection bias does not explain the overly positive opinions. The resulting distribution of evaluations on a five-point scale is presented in Figure 4.13.

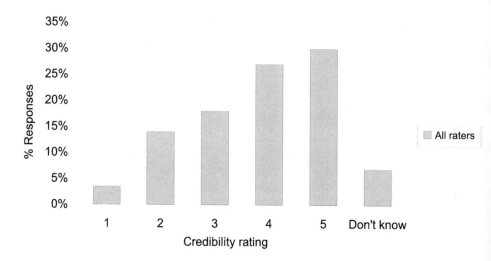

FIGURE 4.13: Distribution of credibility ratings.

What is particularly interesting, in this case, is that additionally a comparison of lay users' ratings and experts' ratings was done. For both groups, 26,9% of pages have equal median credibility scores. The remaining median scores given by the experts are in most cases lower than the corresponding median scores given by the respondents in the main study. This suggests that the Webpages were indeed not that credible and some explanation of the credibility evaluation bias in this case is needed, all the more so when we take into account the fact that the above mentioned bias is observed in several other cases. These include credibility ratings in the WOT system and evaluations of Wikipedia articles [194, 241].

The direct sources of credibility evaluation bias in online sources has not yet been definitely determined. Although definite explanations of this phenomenon still remain unclear, intuitively they should be directly or indirectly connected with characteristics of the individuals making evaluations. Below we describe several possible ways to think about the sources of the cognitive bias in Internet users.

When it comes to description of individual's characteristics influencing Websites' credibility evaluation, there are several factors that presumably should be considered. The most intuitive seem to be personality traits and cognitive heuristics. We dedicated the following two subsections to a brief description of each of these above mentioned variables.

4.4.3.2 Users Characteristics Affecting Credibility Evaluation – Selected Personality Traits

Pursuing the general idea of the Reconcile project (to create a system supporting credibility evaluations), we made an attempt to measure some psychological traits and relate their intensity with a tendency to give certain credibility evaluations. Again, adopting the wisdom of crowds method, we wanted to obtain preliminary results, which in the future would help us to predict credibility misinterpretation tendencies in individuals. Based on those cues, we could implement mechanisms that would moderate evaluations given by users in the final version of the Reconcile system.

To attain this, we conducted a study aimed at determining the relationship between intensity of selected psychological traits and tendencies to misjudge (overestimate or underestimate) Websites' credibility. In February 2013 a team of scientists from Polish-Japanese Institute of Information Technology examined 2046 U.S. adults asking them via Amazon Mechanical Turk to evaluate credibility of presented Websites on a five-point Likert scale. Websites represented different thematic areas and were balanced in terms of credibility level referring to an external credibility index (www.mywot.com). Additionally, personality traits, such as trust, conformity, risk taking and intellect, were measured using the International Personality Item Pool (IPIP)[36].

Based on the median difference between participants' evaluations and the reference external credibility index, they were able to distinguish three disjunc-

---

[36]http://ipip.ori.org/

tive groups of individuals. The group showing a tendency to underestimate Websites' credibility was named under-raters; the group showing the tendency to overestimate Websites was called over-raters. The third group (average-raters) consisted of people giving adequate credibility judgments. Researchers compared mean levels of psychological factors' intensity, which was estimated using graded response model.

Based on the obtained results, we obtained psychological characteristics of individuals showing particular evaluation tendencies. According to our research under-raters can be characterized as a group having the relatively lowest level of trust, low levels of risk-taking and the relatively highest levels of conformity. Over-raters can be characterized as a group obtaining relatively lowest results in intellect scale, low level of risk taking and lowest level of conformity. Average-raters were the group obtaining the relatively highest levels of risk taking, intellect and trust.

In light of the obtained results the IPIP conformity scale might measure experimenters' intentions rather than the general conformity level. As people taking part in the experiment may have suspected that researchers' intentions were to determine what makes a Website not credible, highest results were obtained by under-raters and lowest by over-raters.

Level of trust proved to be a good indicator of specific misjudgment tendencies. As expected, people with relatively lowest trust levels showed a tendency to underestimate Websites' credibility.

As questions in intellect scale are fully declarative (they form a questionnaire not an intelligence test), results in an intellect scale can be interpreted as self-esteem or confidence. So it is justifiable that over-raters achieved lowest scores in this scale. Researchers hypothesize that positive judgments represent in general a "safe option" because as opposed to negative judgments they rarely need to be justified.

Summarizing, the exact pattern of influence psychological factors have on Websites' credibility evaluation is a very promising area of research. Positivity bias (unjustified shift towards positive evaluations) seems to be connected with moderate trust, low risk taking and low self-esteem (intellect). Perhaps adding some new personality factors or including demographic information in further research would help to make this pattern more clear.

4.4.3.3   Users Characteristics Affecting Credibility Evaluation – Cognitive Heuristics

The Internet can be treated as an unlimited source of knowledge. Therefore, finding desirable snippets of information (and its providers) in such an immensity of data needs special approach. Additionally, Internet users' limited time and cognitive capacity make systematic analysis impossible or at least ineffective. Hence, people tend to use cognitive heuristics in evaluating Websites' credibility [179]. It needs to be stressed that using heuristics in making judgments is obviously a phenomenon not restricted to human online behavior. In Metzger and Flanagin [179], based on a literature study and their own research, the authors proposed the categorization of six heuristics in credibility judgment of online contents. There are still many questions that need to be answered when it comes to the use of heuristics in the online environment. First, the structure of heuristics remains unclear. One person can intuitively use several heuristics at a time, but may form perhaps some broader classes or stay in more complex relations.

Persuasive intent heuristic: negative credibility evaluations. This heuristic stems from the observation that people treat biased information as not credible. If people notice that content is presented in a one-sided manner, they are likely to evaluate its credibility negatively. This is mainly connected with advertising messages or propaganda content, and it has the strongest and negative impact, if such piece of persuasive information is not accepted by the receiver.

Expectancy violation heuristic: negative credibility evaluation. This heuristic is connected with negative credibility evaluations. If people consider a Website not coherent with their expectations, they are highly likely to judge its credibility negatively. This might also be associated with unjustified requests for sharing personal data present at a Website.

Self-confirmation heuristic: positive credibility evaluation. This heuristic based on the well-known psychological property that people pay most attention to arguments supporting their beliefs and are more likely to judge the source of those arguments as credible. Such reasoning can also be observed in the online environment.

Consistency heuristic: positive credibility evaluation. This heuristic is linked to the strategy of comparing several sources of information. If the result of such comparison is consistent, the initial source of information may be considered as credible.

Endorsement heuristic: positive credibility evaluation. This heuristic is based on the belief that recommended sources of information are usually credible. The subject making a recommendation might be a respected individual (authority) or a general population of Internet users.

Reputation heuristic: positive or negative credibility evaluation. This heuristic is related to the effect of familiarity. People who somehow recognize an online source of information evaluate it as more credible.

Second, the use of heuristics should be considered in relation to personal characteristics of individuals. For example, it is postulated that the use of heuristics in credibility judgment is moderated by the motivation of an Internet user [180]. Users searching for information that might lead to serious consequences differ in online judgments from those just looking through the Web to kill time.

Moreover, there is some empirical evidence that using heuristics is related to one's expertise level in the topic — people with greater knowledge about the subject tend to rely on heuristic reasoning less than people less knowledgeable [146]. When it comes to frequency of using heuristics, it is linked to many other dimensions, such as personality traits [31] or situational characteristics [180], and from our point of view should be analyzed in the context of the Prominence-Interpretation Theory. The exact relationships seem to be quite complicated and require further studies.

Credibility evaluation shift towards the positive end of the scale is a very intriguing phenomenon, which seems not to be sufficiently described and explained in the present literature. Although there are some hypotheses aiming to solve that puzzle, still much research needs to be done. It must be taken into account that measuring credibility evaluation of online sources is very challenging as it depends strongly on the context and situation of decision making. Also many variables such as different personality traits, demographical factors, cognitive heuristics and many others might interact with one another and bring us closer to the explanation as to why people intuitively give credence to information they find on the Web. This seems plausible only if two main areas of research — one examining Websites' features and the other focusing on personal characteristics — are joined together. From our point of view, this is the only way to examine the wide spectrum of online credibility judgments.

4.5　Aggregation Methods – What Is The Overall Credibility?

4.5.1　How to Measure Credibility

Despite the fact that Web credibility can be considered as a simple feature as well as a multidimensional variable, the most important fact is that credibility is described as a perceived quality, e.g., by Fogg and Tseng [70]. Furthermore, authors have concluded that people may use different strategies for assessing credibility, thus three models of credibility evaluation have been proposed: binary, threshold and spectral. Those strategies are based on the earlier concept of Elaboration Likelihood Model (ELM) of persuasion, introduced by Petty and Cacioppo in the 1980s [209]. According to the ELM model, people can process information through two routes: central and peripheral. In the case where there is little personal involvement with the issue or a lack of motivation to process the information, the peripheral route is chosen. The central route is chosen in the case where there is high personal involvement or high motivation to process the information.

With ELM as a fairly general framework, three different models mentioned above were proposed. The following four elements are used to describe the function of the theoretical credibility: interest in the issue, ability to process information (due to cognitive or situational factors), familiarity with the subject matter and reference points to comparison [70].

According to Fogg and Tseng [70], the simplest approach is the Binary

Evaluation model of credibility which has only two states: user acceptance or rejection. It describes the situation, when users perceive information as either credible or not credible—with no other possibility. This strategy is used if one or more of four elements is present: low interest (in the issue), low ability to process information, little familiarity with the subject or no reference point for comparison.

The opposite strategy is represented in the Spectral Evaluation model of credibility. In this approach all four elements must be present: high interest in the issue, high ability to process the information, high familiarity with the subject and considerable opportunity to compare various sources of information. In this case, there are no black and white categories—users evaluate credibility in shades of gray.

Threshold evaluation of credibility is a mixed approach. As in a binary model, there are upper and lower thresholds for credibility assessment: everything above upper bound is considered as credible, everything below lower threshold is deemed as not credible. But between those two thresholds there is a place for such assessment as: "somewhat credible" or "fairly credible". This is the most common strategy with such elements as moderate interest in the issue, moderate ability to process information, partial familiarity with the subject matter and moderate ability to compare various sources.

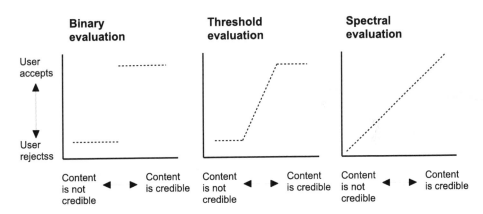

FIGURE 4.14: Credibility evaluation models.

Those three theoretical models (see Figure 4.14) constitute a conceptual framework for evaluating credibility, which should be helpful in further research, leading to better analysis and tools, e.g., scales.

4.5.2 Standard Aggregates

In order to aggregate data, for example, evaluations obtained by the use of a questionnaire about Web credibility, a clear definition of Web credibility is needed. First, a researcher should be sure that the proposed question is understandable and has the same meaning for all responders. Taking into account that Web credibility could be considered subjectively and defined in different ways, a researcher should precisely operationalize indicators of studied notion or phenomenon. Even a given amount of money can be consider subjective and has a different psychological meaning for people. The difference between $1000 and $2000 has different psychological meaning than the difference between $3000 and $3100 [18]. An example of a definition of Web credibility is used by Microsoft in its own research: *A credible webpage is one whose information one can accept as the truth without needing to look elsewhere. If one can accept information on a page as true at face value, then the page is credible; if one needs to go elsewhere to check the validity of the information on the page, then it is less credible* [237].

The construction of a tool, for example, a definition of a question, determines the choice of a scale of measurement. From a statistical point of view, there are five main types of measurement scales:

1. Nominal

2. Ordinal

3. Interval

4. Ratio

5. Absolute

Each scale of measurement has its own properties. Levels of measurement form a cumulative scale, i.e., for instance, an ordinal scale has all the properties of a nominal scale plus properties of an ordinal scale and so on. [18]. Table 4.3 presents the most important characteristics of the mentioned levels of measurement [18, 148].

The usage of particular aggregation method depends on the scale of measurement used. The arithmetic mean, median, maximum and minimum are examples of simple aggregation functions [240]. A weighted or unweighted arithmetic mean is commonly used as an aggregation function, for instance, in recommender systems. But as is shown in Table 4.3, the mean is applicable only in case of an interval or stronger scale. For data measured on an ordinal scale, the median and mode are appropriate. Furthermore, for example, Garcin et al. [77] proved that the mean may not be the most informative measure since it is sensitive to outliers and biases. The arithmetic mean is not a robust statistic. In reality, responders' opinions are often biased and the reviews' distribution is far from the normal distribution. The authors showed that the median, in contrast to the arithmetic mean, handles outliers and malicious attacks, which entails higher recommendation accuracy. The median is a better measure of central tendency for skewed distributions.

Considering the weighted arithmetic mean in terms of aggregating ratings

TABLE 4.3: Variable Measurement Scales and Respective Permissible Operations

Type of scale	Description	Logical/mathematical operations	Permissible operations	Examples
Nominal (qualitative)	A classification with respect to only the measured property is allowed; lack of order of categories; no unit of measurement	$=$ $\neq$	Mode	Gender, nationality, religion
Ordinal	Rank order with respect to the degree is allowed; no unit of measurement; no true zero point	$=$ $\neq$ $>$ $<$	Mode, median	A rating scale, levels of education
Interval	Comparing the exact distances between objects is allowed; a unit of measurement arbitrarily defined; an arbitrarily defined zero point	$=$ $\neq$ $>$ $<$ $+$ $-$ $\dfrac{x_1 - x_2}{x_3 - x_4}$	Mode, median and arithmetic mean	Temperature with the Celsius scale, utility
Ratio	Comparing ratios is allowed; indication of the exact distances between objects is allowed; a unit of measurement arbitrarily defined; a true zero point (non-arbitrary)	$=$ $\neq$ $>$ $<$ $+$ $-$ $\times$ $\div$ $\dfrac{x_1 - x_2}{x_3 - x_4}$ $\dfrac{x_1}{x_2}$	Mode, median, arithmetic and geometric mean	Mass, length, height, incomes, the Kelvin temperature
Absolute	A unit of measurement clearly defined; a true zero point	All numerical operations	All operations	Number of kids

about Web credibility, the definition users' weights is significant. In some systems, for example, in Reconcile, the normalized user's reputation scores may be used as a weight of that user's ratings. In the case of collecting ratings using crowdsourcing or other open methods of collecting data, a researcher may assign weights to responders based on

1. an additional questionnaire regarding user's qualifications,

2. a pretest regarding user's knowledge in a particular field or

3. ratings received as a result of the evaluation of a sample of pages.

In the third example, obtained ratings may be compared with the experts' ratings. The distance between a user's and an expert's ratings may indicate the user's expertise level.

Thinking about levels of measurement and methods of aggregation, let's consider three examples of questions regarding Web credibility:

- Question based on dichotomous scale:

    - If the given Website is credible?
        * Yes
        * No

- Question based on a five-level Likert scale

    - If the given Website is credible?
        * Strongly disagree
        * Disagree
        * Neither agree nor disagree
        * Agree
        * Strongly agree

- Question based on an ordinal scale used in the Reconcile system

    - Rate credibility of this Webpage/Website/text
        * Completely not credible
        * Mostly not credible
        * Somewhat credible, although with major doubt
        * Credible, with some doubt
        * Completely credible

In the first case, a researcher can use all operations applicable for variable containing values equal 0 and 1. In the third example the median is an appropriate measure used to aggregate ratings.

Presented in the second case, a Likert scale is the most controversial since some researchers assume that it is an interval scale and the mean is widely used as a method of aggregating this kind of data. Categories in a Likert scale have a rank order, but it is hard to prove that the intervals between values are equal. Furthermore, data sets based on a Likert scale often have a skewed distribution. In this case, usage of the median or mode as the measure of central tendency is appropriate [106].

4.5.3 Combating Bias – Whose Vote Should Count More?

The dawn of online crowdsourcing platforms, such as Mechanical Turk, has allowed researchers to obtain massive amounts of human-computed data in a short time and at a reasonably low price. This enabled the science community to conduct experiments and researches in fields that were earlier inaccessible, due to high costs [49]. Yet, new solutions brought new problems and the question, whether the quality of an output produced by online workers is satisfying, arose.

On the one hand, the payment for tasks is usually low. The environment, in which workers do their jobs, is far from being supervised. This raises concerns whether Mechanical Turk is robust enough from spammers, who produce high amounts of low quality data [59].

On the other hand, not only attitudes, but also abilities may vary. Even though, as it was shown by the very first crowdsourcing experiment made by Sir Francis Galton [75], crowdsourcing is based on the power of the mediocre, who not necessarily possess precise appraisal; if the incompetent outnumber the skillful, the results tend to be inadequate.

Additionally, there are situations where a "gold standard" is inaccessible, or it would be too expensive to acquire. Imagine a case where workers are expected to appraise the Webs' credibility. There is definitely no gold standard to be expected and while experts are without any doubt better than laymen in such tasks, they are also prone to failure.

Therefore, a reliable method of separating the wheat from the chaff is highly desired. One of the most popular and versatile is the Expectation-Maximization (EM) algorithm introduced in 1977 by Arthur Dempster, Nan Laird, and Donald Rubin, which jointly learns true values and latent variables from observed data [52].

Algorithm can be decomposed into five basic steps [92]:

Step 1. While m=0, make an initial estimate for the set of parameters $\theta(m)$.

Step 2. Expectation step: Using the collected data $\mathbf{x}$ and the set of parameters $\theta(m)$ calculated for all values y_i:

$$p(y_i|\mathbf{x}_i, \theta^{(m)}) \propto p(y_i|\theta^{(m)})p(\mathbf{x}_i|y_i, \theta^{(m)})$$

$$p(y_i|\mathbf{x}_i, \theta^{(m)}) \propto p(y_i) \prod_j p(x_{ij}|y_i, \theta_j^{(m)}))$$

Step 3. Maximization step: With computed posterior probabilities from the Expectation step, create the auxiliary function Q, which is the expected log-likelihood of the observed and hidden variables, conditioned by the parameters.

$$
\begin{aligned}
Q(\theta|\theta^{(m)}) \qquad\qquad &= E[lnp(\mathbf{x}, \mathbf{y}|\theta^{(m)})] \\
&= E[ln \prod_i (p(y_i) \prod_j p(x_{ij}|y_i, \theta_j^{(m)}))] \\
&= \sum_i E[lnp(y_i)] + \sum_{ij} E[lnp(x_{ij}|y_i, \theta_j^{(m)})]
\end{aligned}
$$

Step 4. Using gradient ascent or other optimization techniques, find values for the set of parameters θ, which maximize the Q function. When this is done, θ becomes $\theta(m)$.

Step 5. m := m + 1 and go back to Step 2.

Stop procedure, when values converge:

$$|\theta^{(m)} - \theta^{(m-1)}| < \epsilon,$$

where $\varepsilon > 0$.

This method assures that every iteration will provide parameters and probabilities for labels at least equally as good as the current ones [92]. Still, if most of the workers are misinformed or they consciously give wrong answers, then, without a gold standard, efficiency of this method is limited. Hence, the basic levels of competence and honesty amid workers have to be maintained [49].

One of the first methods using the proposed EM algorithm was made by Dawid and Skene in 1979 [47]. By creating a confusion matrix for every object and every worker, it allows predicting true values for the set of objects and skills of workers involved. However, one of the limitations of this method is that objects used must come from the same kind. In other words, the Dawid and Skene method does not allow using in one experiment objects with different possible responses.

Yet, its relative simplicity made it one of the most common methods implementing the EM algorithm. Currently, one can use Project Troia[37]. Project Troia, in addition to the standard Dawid and Skene model, gives the possibility to evaluate, on the basis of the quality of their output and our cost function, how big the payment should be for each worker.

---

[37]http://project-troia.com/

A more general model was proposed by Paul Ruvolo, Jacob Whitehill, and Javier R. Movellan in 2010 [230]. It not only returns probabilities for true labels and workers' skills, but it also estimates the difficulty of each object to be properly appraised and takes into account the possibility that objects and workers' knowledge might not be independent. For example, people who are enthusiasts of classical music will be better at distinguishing Bach from Mozart than those who are fans of heavy metal. By relaxing the assumption of independence between objects and workers, this model tends to work better than, for example, the Dawid and Skene one [230].

There are also models dedicated to work under certain conditions. An anti-spammers model proposed by Vikas C. Raykar and Shipeng Yu in 2012 [216] is designed to catch automatized workers who give random responses.

4.6 Classifying Credibility Evaluations Using External Web Content Features

In data mining the most general division when it comes to types of research is supervised and unsupervised learning. In the supervised learning (classification), the outcome variable is present and guides the learning process. In the unsupervised approach (clustering), one observes only the features and has no measurements of the outcome [95].

In this section we present a real-life case study of credibility evaluation based on external Webpage features. For this purpose we will focus on supervised learning, an approach for which we have examples, in our case Webpages (represented as a set of features characterizing them) and for each one we know the value of the outcome variable — the credibility of the Webpage. The goal is to train the classifier so that it will be able to predict the new unseen cases based only on the features of the Webpage. Example methods of supervised learning are decision trees, nearest neighbor methods, support vector machines and neural networks.

4.6.1 How We Get Values of Outcome Variable

The value of the outcome variable is known thanks to the experiments carried out on the crowdsourcing platform Amazon Mechanical Turk for the Reconcile project. The wider description of the data can be found in Section 4.3.3.

4.6.2   Motivation for Building a Feature-Based Classifier of Webpages Credibility

The main motivation is to build an algorithm (or rather a set of them), which would be able to automatically determine whether an unknown Webpage (unseen case in terms of classification) is credible. Such a tool could then be used in many different areas and applications, for example, a Web browser plugin, parental control tools, spam detection, increasing the quality of search results in terms of reliability. All that could have big social impact on everyday Internet users, as mentioned in the introduction to this chapter. There is a pressing need for automated tools for credibility evaluation because of the rapid pace of Web growth, which implies intractability (mostly because of high cost and lack of appropriate incentives for volunteers) of manual evaluations of credibility made by humans.

4.6.3   Classification of Web Pages Credibility – Related Work

It is worth mentioning the research by Olteanu et al. [198] concerning classifying credibility based on a robust set of content-based and social-based Webpage features and the research by Sondhi, Vydiswaran, and Zhai [247] dealing with a specific niche — Webpages from the medical domain. The first is more general, based on a dataset from Microsoft, which is described in detail by Schwarz and Morris [237]. This set consists of 1000 URLs with credibility evaluated on a 5-point Likert scale (score of 1 for "very noncredible" and 5 for "very credible") using the definition of credibility described by Schwarz and Morris [237]. Olteanu builds a binary classifier and achieves an accuracy of 75% (with precision and recall on the level of 70%).

The authors dealing with reliability of Webpages in medical domain have prediction accuracies of over 80%. In the case of learning a classifier, the gold standard dataset constructed by experts was used. The data set is relatively small: 360 Webpages (with 1:1 proportion of reliable and unreliable). Because of the nature of the topic required a reasonable amount of expertise in understanding the criteria and content, the authors decided not to use Amazon Mechanical Turk for a broad crowdsourcing acquirement of evaluations. The feature set consisted of link-based, commercial, PageRank, presentation and word features.

In both papers the authors focused only on a binary classification. The authors dealing with medical domain emphasized the positive aspects of such a dichotomous approach: ease of judgments, no problem with definition of intermediate class and no problem with evaluation of criteria by themselves if manually defined.

Both papers do not consider the issue of controversy.

4.6.4 Dealing with Controversy Problem

As mentioned in Section 4.4, the evaluations collected in the Amazon Mechanical Turk experiment are subjective and for some of Websites it is not even possible to aggregate the ratings to one reasonable evaluation because of the Web page controversial nature, which implies a high diversity of ratings. Before we start to assign Websites to one of the credibility classes, we want to determine whether the given Website is controversial. Having a learning set with many ratings per page, one can distinguish whether the page is controversial, based on standard deviation of the given ratings or, in a more sophisticated way, using a Leik measure. However we need to build a classifier, which will be using only Webpage features (no human ratings will be available) to discover controversy, which will help to eliminate cases of sites which are not appropriate for further classification in terms of credibility. For training such a classifier one of the computed consensus measures on ratings (such as Tastle, Eijk or Leik) can be used or the crowdsourcing workers can be asked to vote if the Webpage is or is not controversial. Figures 4.15 and 4.16 show the relation between Leik and standard deviation of ratings.

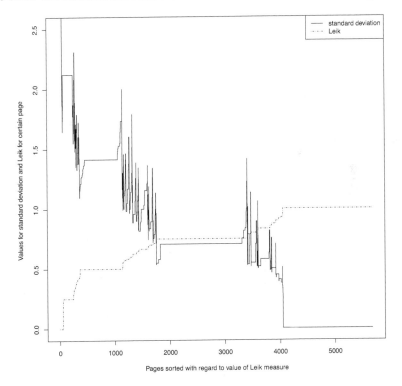

FIGURE 4.15: Leik sorted ascending and corresponding value of standard deviation.

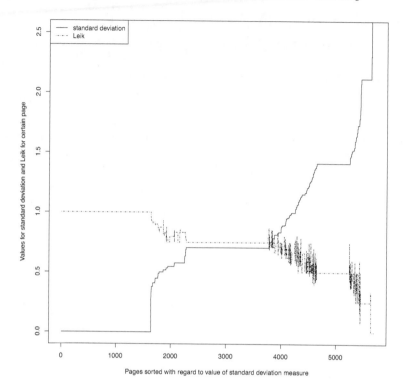

FIGURE 4.16: Standard deviation sorted ascending and corresponding value of Leik.

Rejection of controversy cases significantly improves the accuracy and recall of classification.

4.6.5   Aggregation of Evaluations

While preparing the training set for our classifier, we faced the problem of how to aggregate the results. We tested the following ways of aggregation: minimum, median and average of ratings, and we also tested the labels provided by Project Troia, a system for evaluating the quality of completed crowdsourcing tasks using the Expectation-Maximization algorithm. The system and the algorithm are described in Section 4.5.

In the research, only the pages with two and more evaluations per page were taken into consideration, and in some scenarios even a higher number of evaluations was demanded. Figure 4.17 presents the distribution of the number of ratings received.

Increasing this threshold causes a better accuracy in the model, which will confirm the hypothesis about the wisdom of crowds, but it also has a

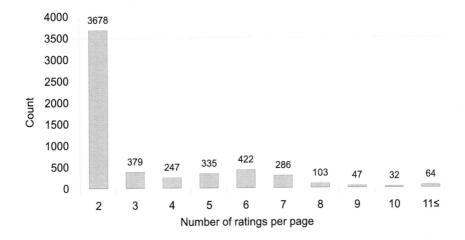

FIGURE 4.17: Count of pages with certain number of ratings per page.

drawback: lessen in the number of observations and risk of overfitting the model. The other observation is that on the one hand we want to learn from many ratings, but on the other, the higher the number of ratings is, the smaller the consensus, which is shown in Figure 4.18.

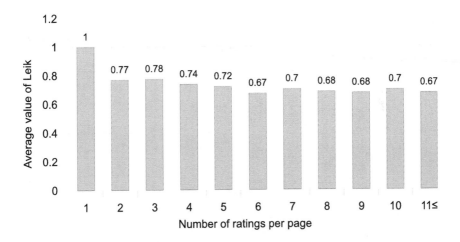

FIGURE 4.18: The average value of Leik.

4.6.6 Features

We extracted features automatically from Websites and from APIs of 3rd party services. The starting point for selection of features was the article by

Olteanu et al. [198]. We divided the features into two big groups: content-based and social-based. Below we present only the statistically important features, which were used in the construction of predictive models.

1. Content-based

    *   RB — Number of adverbs
    *   css_definitions — Number of Webpage CSS style definitions
    *   document_url — Document domain (.edu, .com, etc.)
    *   exclamations — Number of exclamation marks ("!") in text
    *   questions — Number of question marks ("?") in text
    *   smog — Statistical measures of text readability

2. Social-based

    *   tweets — Number of Tweets mentioning a Webpage URL
    *   fb_clicks — Number of Facebook clicks for a Webpage URL
    *   fb_comments — Number of Facebook comments for a Webpage URL
    *   fb_likes — Number of Facebook likes for a Webpage URL
    *   fb_shares — Number of Facebook shares for a Webpage URL
    *   fb_total — Total Facebook shares, likes, comments and clicks
    *   delicious_bookmarks — Number of Delicious[38] bookmarks for a Webpage URL
    *   alexa_linksin — Number of Website linkings estimated by Alexa[39]
    *   alexa_rank — Alexa rank
    *   bitly_clicks — Number of Bitly[40] short URL clicks for a Webpage
    *   bitly_referrers — Number of Websites having Bitly short URL for a Webpage
    *   page_rank — Google PageRank

Table 4.4 presents the correlation matrix showing the level of correlation between features as well as the outcome variable.

---

[38]https://delicious.com
[39]http://www.alexa.com/
[40]https://bitly.com

TABLE 4.4: Correlation Matrix between Explanatory Variables and Different Variant for Outcome Variable.

	AVERAGE	MEDIAN	MINIMUM	TROIA	MODE
alexa_linksin	0.043	0.043	-0.008	0.006	0.063
alexa_rank	-0.071	-0.083	-0.046	-0.037	-0.091
bitly_clicks	0.002	0.015	-0.014	0.010	0.019
bitly_referrers	0.041	0.048	0.004	0.033	0.063
css_definitions	0.168	0.157	0.112	0.088	0.167
delicious_bookmarks	0.034	0.020	0.016	0.005	0.032
document_url	-0.003	0.000	0.004	-0.002	0.004
exclamations	-0.063	-0.045	-0.067	-0.023	-0.066
fb_clicks	0.012	0.016	-0.016	0.038	0.023
fb_comments	0.026	0.029	0.008	0.029	0.039
fb_likes	0.015	0.020	0.000	0.029	0.030
fb_shares	0.007	0.006	-0.001	0.027	0.014
fb_total	0.017	0.020	0.002	0.032	0.030
page_rank	0.285	0.243	0.209	0.130	0.265
questions	-0.022	-0.019	-0.027	-0.002	-0.013
RB	0.019	0.015	-0.007	0.007	0.023
smog	0.078	0.066	0.071	0.033	0.068
tweets	0.001	0.013	-0.014	0.007	0.016

4.6.7 Results of Experiments with Building of Classifier Determining whether a Webpage Is Highly Credible (HC), Neutral (N) or Highly Not Credible (HNC).

As mentioned in Section 4.6.1, the workers were evaluating the credibility of Webpages on the 5-level scale, which implies the concern about the bias of experimental results.

The way to deal with bias, in ratings given by Amazon Mechanical Turk, should be the use of tools such as Project Troia, supported with "gold" data with ratings given by experts for precisely selected Webpages. However, because the "gold" data for Reconcile credibility corpus is yet collected, the algorithm was used without such information. Note that the results should get better when the "gold" data is provided.

Since up to now the results of classification with outcome variable set to the Troia label have been very poor, we chose the median for further proceeding.

Before we start aggregating the results, we reject all the observations of Webpages, in the way that a single Webpage:

1. has fewer than 5 evaluations or

2. the Leik of the evaluations for that Webpage is less than 0.1.

We end up with a set of 1288 observations.

For now, in the aggregation process we are using as the output variable the median discretized in 3 classes as follows:

The set of features used for modeling is the same as noted in Section 4.6.6.

TABLE 4.5: Median Rating Recoding

Highly Not Credible (HNC)	rating = 1	1% of all observations
Neutral (N)	rating = 2 , 3 or 4	59% of all observations
Highly Credible (HC)	rating = 5	40% of all observations

For the process of building different models and preparing data, the data mining software RapidMiner[41] in the Community Edition version was used.

As the data mining technique, decision trees were selected.

When using the decision trees algorithm, we use information gain as the criteria to construct the tree. At every step, the entropy based information gain of every attribute is calculated, and the one with the maximum information gain is selected for tree building. Prepruning is also applied.

We obtained the following results for running the algorithm with prepruning and allowing maximal depth of tree to be equal to 8 and the minimum information gain set to be at least 0.028.

The Accuracy is 69.7% and the weighted precision and recall are, respectively, 52.3% and 49.2%. The other configurations of different max depth or gain ratio are in some cases better in terms of accuracy, but because of the phenomenon known as accuracy paradox, we tended to use precision and recall, and because they are often both contradictory for comparison of different models, we used F-measure calculated as

$$F = 2 * precision * recall/(precision + recall)$$

The selected parameters for the decision tree were best in terms of F-measure, which was equal to 0.51. The confusion matrix for the mentioned model is presented in Table 4.6.

TABLE 4.6: Confusion Matrix for Three-Class Model

	True HNC	True N	True HC	Class precision
Pred. HNC	2	5	3	20.00%
Pred. N	13	607	224	71.92%
Pred. HC	3	142	289	66.59%
Class recall	11.11%	80.50%	56.01%	

Figure 4.19 shows the history of splits in the form of the tree with split thresholds on the edges.

For information, further we present the textual version of the decision tree with the number of cases that are classified.

---

[41] http://rapidminer.com/

FIGURE 4.19: Tree view for decision tree with max depth=8 and minimum information gain=0.028.

```
page_rank > 3.500
| alexa_rank > 433258.500
| | delicious_bookmarks > 3.500: N {HNC=0, N=10, HC=0}
| | delicious_bookmarks = 3.500
| | | css_definitions > 99
| | | | css_definitions > 385.500
| | | | | alexa_rank > 4242791: N {HNC=0, N=3, HC=0}
| | | | | alexa_rank = 4242791
| | | | | | alexa_rank > 1064326: HC {HNC=0, N=0, HC=6}
| | | | | | alexa_rank = 1064326: N {HNC=0, N=6, HC=3}
| | | | css_definitions = 385.500
| | | | | fb_likes > 18.500: N {HNC=0, N=2, HC=1}
| | | | | fb_likes = 18.500
| | | | | | RB > 54: HC {HNC=0, N=0, HC=5}
| | | | | | RB = 54: HNC {HNC=4, N=0, HC=3}
| | | css_definitions = 99
| | | | fb_clicks > 0.500: HC {HNC=0, N=0, HC=3}
| | | | fb_clicks = 0.500
| | | | | exclamations > 0.500
| | | | | | RB > 30: HC {HNC=0, N=0, HC=3}
| | | | | | RB = 30: N {HNC=0, N=3, HC=2}
```

```
| | | | | exclamations = 0.500: N {HNC=0, N=12, HC=1}
| alexa_rank = 433258.500: HC {HNC=0, N=114, HC=267}
page_rank = 3.500
| css_definitions > 19.500: N {HNC=5, N=463, HC=199}
| css_definitions = 19.500
| | alexa_rank > 20479.500
| | | alexa_linksin > 54
| | | | RB > 215
| | | | | fb_shares > 0.500: HC {HNC=0, N=1, HC=2}
| | | | | fb_shares = 0.500: N {HNC=1, N=2, HC=0}
| | | | RB = 215: N {HNC=7, N=91, HC=2}
| | | alexa_linksin = 54
| | | | questions > 3.500: HC {HNC=1, N=0, HC=2}
| | | | questions = 3.500
| | | | | smog > 16.513
| | | | | | alexa_rank > 5072191: HC {HNC=0, N=0, HC=4}
| | | | | | alexa_rank = 5072191: N {HNC=0, N=2, HC=0}
| | | | | smog = 16.513: N {HNC=0, N=17, HC=1}
| | alexa_rank = 20479.500
| | | bitly_referrers > 9.500: HC {HNC=0, N=0, HC=3}
| | | bitly_referrers = 9.500
| | | | RB > 35.500
| | | | | css_definitions > 0.500: N {HNC=0, N=2, HC=1}
| | | | | css_definitions = 0.500: HC {HNC=0, N=0, HC=3}
| | | | RB = 35.500
| | | | | alexa_linksin > 2.500: N {HNC=0, N=25, HC=3}
| | | | | alexa_linksin = 2.500: HC {HNC=0, N=1, HC=2}
```

The analysis of the confusion matrix shows that this model has rather limited predictive power. It is hard to find a benchmark of such models based on data collected from crowdsourcing, because most of the publications deal only with binary classification.

4.6.8   Results of Experiments with Build of Binary Classifier Determining whether Webpage Is Credible or Not

In order to have comparable conditions in experimenting with the binary classifier, we also used a dataset with the same examples as in the three-class experiment (i.e., only the cases of Web pages with minimal number of 5 ratings, and controversy threshold measured by Leik equal to 0.1). We chose a threshold for discretization: if median > 4, the Webpage is credible; otherwise it is not. We ended up with 516 cases of credible Webpages and 772 of not credible.

This time for the method of aggregation we chose median, we allowed the decision tree to have the maximal depth of 10, and we set the minimum gain ratio to 0.03. We obtained 71.4% accuracy, by precision on the level of 70.5% and recall on the level of 68.55%. The exact performance is shown in the confusion matrix in Table 4.7

The decision tree with the history of splits and thresholds, as well the exact number of cases classified, is shown in Figure 4.20. A textual description of the classification is provided after the figure.

TABLE 4.7: Confusion Matrix for Binary Classifier

	True NOT_CREDIBLE	True CREDIBLE	Class Precision
Predicted NOT_CREDIBLE	641	237	73.01%
Predicted CREDIBLE	131	279	68.05%
Class recall	83.03%	54.07%	

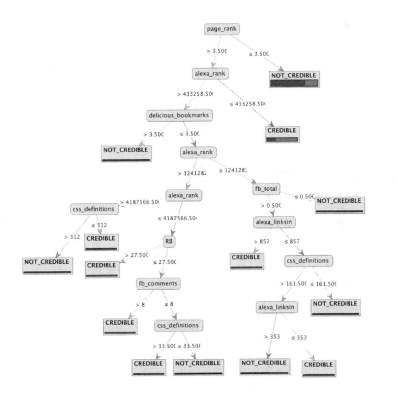

FIGURE 4.20: Tree view for decision tree with max depth=10 and minimum information gain=0.03.

```
page_rank > 3.500
|  alexa_rank > 433258.500
|  |  delicious_bookmarks > 3.500: NOT_CRED {NOT_CRED=9, CRED=0}
|  |  delicious_bookmarks = 3.500
|  |  |  alexa_rank > 1241282
|  |  |  |  alexa_rank > 4187566.500
|  |  |  |  |  css_definitions > 312: NOT_CRED {NOT_CRED=3, CRED=0}
|  |  |  |  |  css_definitions = 312: CRED {NOT_CRED=1, CRED=1}
|  |  |  |  alexa_rank = 4187566.500
|  |  |  |  |  RB > 27.500: CRED {NOT_CRED=0, CRED=8}
|  |  |  |  |  RB = 27.500
|  |  |  |  |  |  fb_comments > 8: CRED {NOT_CRED=0, CRED=4}
|  |  |  |  |  |  fb_comments = 8
|  |  |  |  |  |  |  css_definitions > 33.500: CRED {NOT_CRED=1, CRED=3}
|  |  |  |  |  |  |  css_definitions = 33.500: NOT_CRED {NOT_CRED=4, CRED=0}
|  |  |  alexa_rank = 1241282
|  |  |  |  fb_total > 0.500
|  |  |  |  |  alexa_linksin > 857: CRED {NOT_CRED=0, CRED=4}
|  |  |  |  |  alexa_linksin = 857
|  |  |  |  |  |  css_definitions > 161.500
|  |  |  |  |  |  |  alexa_linksin > 353: NOT_CRED {NOT_CRED=3, CRED=2}
|  |  |  |  |  |  |  alexa_linksin = 353: CRED {NOT_CRED=0, CRED=2}
|  |  |  |  |  |  css_definitions = 161.500: NOT_CRED {NOT_CRED=6, CRED=0}
|  |  |  |  fb_total = 0.500: NOT_CRED {NOT_CRED=11, CRED=0}
|  alexa_rank = 433258.500: CRED {NOT_CRED=108, CRED=250}
page_rank = 3.500: NOT_CRED {NOT_CRED=574, CRED=208}
```

4.6.9   Results of Experiments with Build of Binary Classifier of Controversy

As mentioned in Section 4.6.5, for a fully equipped model we need a classifier of controversy. The k-Nearest Neighbors method was applied to all the features mentioned in Section 4.6.7, which were normalized due to the numerical nature of the k-NN method. The level of $k = 10$ was determined experimentally. Using 15-fold cross-validation we obtained precision of 77% and weighted mean recall of 63%. Webpages having at least 4 ratings were the examples selected for training the classifier, and examples with standard deviations of ratings higher than 0.75 were labeled as controversial.

4.6.10   Summary and Improvement Suggestions

The introduction of concept of controversy improves the results of classification and provides an opportunity to possibly develop other methods of this kind of evaluation.

The problem of bias is surely affecting the classification, but to be able to deal with it consciously, further research is needed, for example, with the use of Project Troia, while providing "gold" labels.

Using only accuracy as a measure of the model's quality is insufficient and misleading. While considering precision and recall, one can control the predictive power of classifiers, avoiding better accuracy paradox. Depending on what the feature applications of the classifier should be, one can consider a choice of model with a better recall in prediction of a certain class, e.g., if the mechanism should be used in a Web browser plugin warning less advanced Internet users about the potentially untrustworthy content, then the recall for the class "not credible" should be as high as possible.

Chapter 5

Trust-Aware Recommender Systems

Recommender systems are an effective solution to the information overload problem, especially in the online world where we are constantly faced with inordinately many choices. These systems try to find the items such as books or movies that match best with users' preferences. Based on the different approaches to finding the items of interest to users, we can classify the recommender systems into three major groups. First, content-based recommender systems use content information to make a recommendation. For example, such systems might recommend a romantic movie to a user who showed interest in romantic movies in her profile. Second, collaborative filtering recommender systems rely only on the past behavior of the users such as their previous transactions or ratings. By comparing this information, a collaborative filtering recommender system finds new items to users. In order to address the cold-start problem and fend off various types of attacks, the third class of recommender systems, namely *trust-aware recommender systems*, is proposed. These systems use social media and trust information to make a recommendation, which is shown to be promising in improving the accuracy of the recommendations. In this chapter, we give an overview of state-of-the-art recommender systems with a focus on trust-aware recommender systems. In particular, we describe the ways that trust information can help to improve the quality of the recommendations. In the rest of the chapter, we introduce *recommender systems*, then *trust in social media*, and next *trust-aware recommender systems*.

5.1   Recommender Systems

With the development of Web 2.0, information has increased at an unprecedented rate which aggravates the severity of the information overload problem for online users. For example, a search for "smartphone" returns 1,664,253 results in Amazon[1] products or a search for "best movies to watch" in Google videos[2] returns about 219,000,000 results. Due to the information overload problem, the decision-making process becomes perplexing when one is exposed to excessive information [2,97,220,234]. Therefore, with the rapidly growing amount of available information and digital items on the Web, it is necessary to use tools to filter this information in order to find items that are more likely to be of interest to the users.

One can use search engines to overcome the information overload problem. In this case, the user has to refine the search terms or pick more specific query terms to narrow down the results. Another solution to overcome the information overload problem is to use top-k recommendations. In this ap-

---

[1]http://www.amazon.com
[2]https://www.google.com/#q=best+movies+to+watch&safe=active&tbm=vid

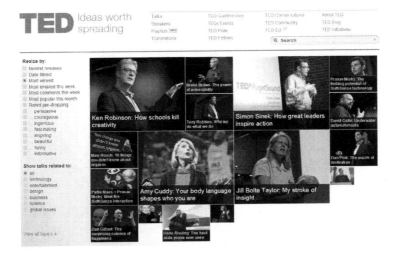

FIGURE 5.1: Ted.com uses a top-k item recommendation approach to rank items.

proach, the system keeps a list of the most popular items and utilizes the list to recommend items to the user. For example, Ted[3] is a Website that uses this technique to recommend items to users. It can be seen in Figure 5.1; users can sort items based on the different approaches such as overall popularity (*most viewed*), popularity in the past week (*most emailed this week*), or popularity in the past month (*most popular this month*) among others. Similar to search engines, top-k items are not usually customized based on users' preferences and interest. In particulate, a top-k-item system returns the same list of items to people with different preferences. Therefore, customization is the major problem associated with these two approaches.

Recommender systems are introduced to tackle the information overload and the customization problem. Recommender systems are a subclass of information filtering systems that consider users' preferences and recommend items that match with these preferences and interests. These systems have become extremely common in recent years and are applied in a variety of applications including recommending products, social links, and digital items. The most popular ones are probably movies, music, news, books, and products in general [97, 122, 234, 240, 266]. Further, recommender systems are frequently used on recommending social links such as recommending people to follow on Twitter, befriend on social networks or dating sites [159, 261]. Furthermore, these systems are also used to accurately estimate the degree to which a particular user (from now on termed the *target user*) will like a particular item (the *target item*) [273].

Based on the type of data that recommender systems use, we can classify

---

[3]http://www.ted.com/

them into two major classes: *content-based* and *collaborative filtering*-based recommender systems [240, 295]. Content-based recommendation systems use items' features and characteristics to rank the items based on the user's preferences. Collaborative filtering recommendation systems rely on the user's past behavior, e.g., purchases or ratings, to find similar users or items and utilize this information in order to find the items of interests to the user.

In general, recommender systems are utility functions that predict the rating of item i from the item set I for user u from the user set U in the form of $U \times I \to R$, where r_{ui} is the rating of the item i for the given user u. The task of recommender systems is to predict user u's rating for the given item i for which r_{ui} is unknown and use $\hat{r}_{ui}$ to represent the predicted rating. The ratings, $r_{u,i}$, can be any real number but often ratings are integers in the range $[1, 5]$. We use $\mathbf{R}$ to show all of the ratings. In real-world recommender systems, only a few users rate the items of interests (this number for many recommender system is less than 1%). Matrix 5.1 shows an example of a rating matrix with missing values. The goal of recommender systems is to predict these missing values.

$$\mathbf{R} = \begin{bmatrix} & 5 & 2 & & 3 & \\ 4 & & & 3 & & 4 \\ & & 2 & & & 2 \\ 5 & & & 3 & & \\ & 5 & 5 & & & 3 \end{bmatrix} \tag{5.1}$$

5.1.1 Content-Based Recommendation

Content-based recommender systems use items' and users' features to create a profile for each item or user. For example, a movie profile might include attributes such as gender, participating actors, director, and office box popularity. User profile includes demographic information and users' interests [124]. These systems use supervised machine learning to induce a classifier that can discriminate among items likely to be of interest to the user and those likely to be uninteresting [15, 187, 206]. The recommender recommends an item to a user based on a description of the item and a profile of the user's interests. Algorithm 1 shows the main steps of a content-based recommendation.

We usually use *vector space model* to represent users' and items' features. In this model, every item or user is represented as a vector.

$$i = (t_1, t_2, ..., t_n) \tag{5.2}$$

where t_j is the frequency of term j in item i. To model users or items more accurately, instead of frequency we can use *tf-idf* (term frequency - inverse document frequency) which can be calculated as follows:

$$tf_{t,i} = \frac{f_{t,i}}{max\{f_{z,i} : z \in i\}} \quad idf_t = \log \frac{N}{n_t} \tag{5.3}$$

Algorithm 1 Content-based recommendation

1: Describe the items that may be recommended.
2: Create a profile of the user that describes the types of items the user likes
3: Compare items to the user profile to determine what to recommend. The profile is often created and updated automatically in response to feedback on the desirability of items that have been presented to the user.

$$w_{t,i} = tf_{t,i} \times idf_t \qquad (5.4)$$

where $f_{t,i}$ is the frequency of term t in item i, $max\{f_{z,i} : z \in i\}$ is the maximum term frequency in item i, N is the total number of items, n_t is the number of items where term t appears $tf_{t,i}$ denotes the frequency of term t in item i and idf_t denotes the inverse document frequency of term t, which inversely correlates with the number of items, in which term t appears in their descriptions. The similarity between user u and item i can be calculated using Equation 5.5.

$$sim(u,i) = \frac{\sum_{t \in T} w_{t,u} w_{t,i}}{\sqrt{\sum_{t \in T} w_{t,u}^2} \sqrt{\sum_{t \in T} w_{t,i}^2}} \qquad (5.5)$$

where T indicates the set of terms that appeared in item and user description.

5.1.2 Collaborative Filtering (CF)

Collaborative filtering is the process of filtering the information or patterns using techniques involving collaboration among multiple agents, viewpoints, data sources, etc. [265]. Collaborative filtering systems use the users' past behaviors, and recommend items that match their taste. Collaborative filtering recommender systems can be classified into *memory-based* and *model-based* collaborative filtering. In the memory-based approach we predict the missing ratings based on similarity between users or items. In model-based approach, we use given user–item ratings to construct a model and use the model to predict missing ratings. We'll give a detailed description of these two approaches in the following sections. The main advantage of this method is that the recommender system does not need to have any information about the users and content of the items to recommend. User–item ratings are the only information the system needs to operate. The following are assumptions for collaborative filtering systems [295]:

- Users with similar ratings on some items are more likely to have similar ratings on future items, and

- Items with similar ratings in the past are more likely to have similar ratings in the future.

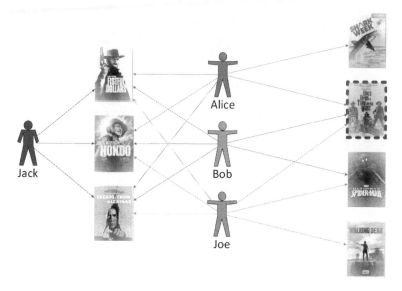

FIGURE 5.2: User-based collaborative filtering.

Figure 5.2 illustrates this approach for a small set of users and movies. The goal is recommending a new movie to Jack. In the first step, the system finds three other users that have similar movie taste as Jack's. In the next step it looks for other movies that these users have liked. All three of them liked "Once Upon a Time in the West", and two of them liked "Spider Man". Therefore, the top recommendation would be "Once Upon a Time in the West".

5.1.2.1 Memory-Based Collaborative Filtering

In a memory-based approach, the recommender system aims to predict the missing ratings based on either similarity between users or similarity between items [20,141,207]. The former is built upon the hypothesis that similar users have similar tastes. Hence, to make a reasonable recommendation, it finds similar users, then uses these users' tastes to make a recommendation for the target user. The second approach is built upon the consistency of a user's taste. If a user liked a product, she will like similar products as well. In both approaches, the recommender system takes two steps to recommend an item to the user. First, it calculates the similarity between users or the similarity between items. Then, it uses the most similar users or items to make its recommendation. Collaborative filtering uses the rating information to compute user–user or item–item similarity. *Cosine similarity* and the *Pearson correlation coefficients* are two of the most frequently used similarity measures in collaborative filtering. Given two users, u and v, and their ratings, the cosine similarity is represented as the following using a dot product and its magni-

---

Algorithm 2 Item-based collaborative filtering

1: Build an item–item matrix determining relationships between pairs of items

2: Using the matrix and the data on the current user, infer his taste

---

tude:

$$similarity(u, v) = \cos(u, v) = \frac{u \cdot v}{\|u\| \|v\|} = \frac{\sum_i r_{u,i} \times r_{v,i}}{\sqrt{\sum_i (r_{u,i})^2} \times \sqrt{\sum_i (r_{v,i})^2}}, \quad (5.6)$$

where $r_{u,i}$ indicates user u's rating for item i. Pearson's correlation coefficient between two variables is defined as the covariance of the two variables divided by the product of their standard deviations:

$$sim(u, v) = \frac{\sum_i (r_{u,i} - \bar{r}_u)(r_{v,i} - \bar{r}_v)}{\sqrt{\sum_i (r_{u,i} - \bar{r}_u)^2} \sqrt{\sum_i (r_{v,i} - \bar{r}_v)^2}}, \quad (5.7)$$

where $\bar{r}_u$ denotes the average rating for user u.

Item-based CF

Item-based collaborative filtering proceeds in an item-centric manner. The recommender system builds an item–item matrix determining relationships between pairs of items; then, it infers the tastes of the current user by examining the matrix and matching that user's data [147, 233]. This method was invented and first used by Amazon (*users who bought x also bought y*). A high level algorithm for item-based collaborative filtering is described in Algorithm 2 [141]. We use Equation 5.8 to estimate $\hat{r}_{u,i}$, denoting the user u's rating for item i:

$$\hat{r}_{u,i} = \bar{r}_i + \alpha \sum_{j \in N(i)} (sim(i, j) \times (r_{u,j} - \bar{r}_j)), \quad (5.8)$$

where i and j denote the items, $\bar{r}_i$ denotes the mean rating of item i, and α is a normalization factor. Many popular Web-based service providers such as Amazon and YouTube use item-based collaborative filtering to recommend items.

User-based CF

In a user-based collaborative filtering approach, the recommender ranks users based on the similarity among them, and uses suggestions provided by most similar users to recommend new items [278]. The user-based approach of collaborative filtering systems is not as preferred as an item-based approach due to the instability in the relationships between users. For a system which handles a large user base, even the smallest change in the user data is likely to reset the entire group of similar users. We use Equation 5.9 to predict user u's rating for item i, $\hat{r}_{u,i}$:

$$\hat{r}_{u,i} = \bar{r}_u + \alpha \sum_{v \in N(u)} (sim(u,v) \times (r_{v,i} - \bar{r}_v)), \tag{5.9}$$

where $r_{v,i}$ is the observed rating of user v for item i, $\bar{r}_u$ is the mean rating of user u, $\hat{r}_{u,i}$ is the predicted rating of user u for item i, $N(u)$ is the set of users similar to u, $sim(u,v)$ is the similarity between users u and v, and α is normalization factor.

5.1.2.2   Model-Based Collaborative Filtering

Unlike memory-based collaborative filtering which uses the similarity between users and items to predict missing ratings, model-based collaborative filtering assumes that an underlying model governs the way users rate the items [295]. In this approach, we aim to learn that model and then use it to predict the missing ratings [207]. Model-based approaches use data mining and machine learning algorithms to find patterns from user ratings. There are many model-based collaborative filtering algorithms, including Bayesian networks, factorization and latent semantic models such as singular value decomposition, and Markov decision process-based models [99, 183, 252]. In this section we introduce matrix factorization model-based collaborative filtering.

Factorization-based CF models assume that a few latent patterns influence user rating behaviors and perform a low-rank matrix factorization on the user–item rating matrix. Let $\mathbf{U}_i \in \mathbb{R}^K$ and $\mathbf{V}_j \in \mathbb{R}^K$ be the user preference vector for u_i and the item characteristic vector for v_j, respectively, where K is the number of latent factors. Factorization-based collaborative filtering models solve the following problem:

$$\min_{\mathbf{U},\mathbf{V}} \sum_{i=1}^{n} \sum_{j=1}^{m} \mathbf{W}_{ij} (\mathbf{R}_{ij} - \mathbf{U}_i \mathbf{V}_j^\top)^2 + \alpha(\|\mathbf{U}\|_F^2 + \|\mathbf{V}\|_F^2), \tag{5.10}$$

where $\mathbf{U} = [\mathbf{U}_1^\top, \mathbf{U}_2^\top, \ldots, \mathbf{U}_n^\top]^\top \in \mathbb{R}^{n \times k}$ and $\mathbf{V} = [\mathbf{V}_1^\top, \mathbf{V}_2^\top, \ldots, \mathbf{V}_m^\top]^\top \in \mathbb{R}^{m \times K}$. K is the number of latent factors (patterns), which is usually determined via cross-validation. The term $\alpha(\|\mathbf{U}\|_F^2 + \|\mathbf{V}\|_F^2)$ is introduced to avoid overfitting, controlled by the parameter α. $\mathbf{W} \in \mathbb{R}^{n \times m}$ is a weight matrix where $\mathbf{W}_{ij}$ is the weight for the rating of user u_i for item v_j. A common way to set $\mathbf{W}$ is $\mathbf{W}_{ij} = 1$ if $\mathbf{R}_{ij} \neq 0$. The weight matrix $\mathbf{W}$ can also be used to handle the implicit feedback and encode side information such as user click behaviors [62], similarity between users and items [144, 201], quality of reviews [215], user reputation [260] and temporal dynamics [124].

5.1.3   Hybrid Recommendation

Each of the two prominent recommendation approaches in this section use a different source of information to generate the recommendation. One method to improve the quality of the recommendation is to use a combination of the approaches [11, 26]. In this section, we introduce methods for combining the results from other systems to improve the quality of recommendation.

- *Monolithic hybridization.* In this approach, there is only one recommendation system. This recommendation system uses different sources and approaches to make the recommendation. So this case is a modification in the base behavior of recommender algorithms to make another one that can use the advantages of all of the other algorithms.

- *Parallelized hybridization.* In this approach we use more than one recommender algorithm. The hybridization process gets the results of recommender systems, combines them and generates the final recommendation. We can use different techniques to combine the results of other recommender systems, such as *mixed* (a union of results from all recommender systems), *weighted* (a weighted combination of the results), *switching* (using results from specific recommender systems for specific tasks) and *majority voting.*

- *Pipeline hybridization.* In this approach, we also use more than one recommender system, and we put the recommender systems in a pipeline. The result of one recommender is the input for another one. The earlier recommender can make a model of the input and pass it to the next recommender system or can generate a list of recommendations to be used by the next recommender system.

5.1.4   Evaluating Recommender Systems

There are many different recommender systems. To find the best one for a special task, we need to evaluate the results. Recommender systems can be evaluated based on the *accuracy of prediction, accuracy of classification, accuracy of ranks* and *acceptance level* [97, 122].

Evaluating the Accuracy of Prediction To evaluate the accuracy, we need to have prediction data as well as ground truth. We get a subset of ratings as the test set, then run the algorithm and get the recommendations for missing user–item values. We can use *Mean Absolute Error (MAE)* or *Root Mean Square Error (RMSE)* to evaluate the accuracy of prediction [107].

$$MAE = \frac{1}{n} \times \sum_{(u,i)\in Testset} |\hat{r}_{u,i} - r_{u,i}| \qquad (5.11)$$

MAE is the quantity used to measure the average deviation between a recommender system's output $\hat{r}_{u,i}$ and the actual rating values $r_{u,i}$ for all evaluated users from the test set of size n. Comparing with MAE, RMSE puts more stress on the deviations.

$$RMSE = \sqrt{\frac{\sum\limits_{(u,i) \in Testset} (\hat{r}_{u,i} - r_{u,i})^2}{n}} \qquad (5.12)$$

To remove the effect of the different ranges of ratings, we can normalize MAE by dividing it by $|r_{max} - r_{min}|$. The new measure is called *normalized MAE (NMAE)* and is calculated as follows.

$$NMAE = \frac{MAE}{r_{max} - r_{min}} \qquad (5.13)$$

Evaluating the Accuracy of Classification We can use precision and recall to evaluate the algorithm for top-k recommendation. *Precision* is the number of correct results divided by the number of all returned results and tells us how accurate the prediction is.

$$Precision = \frac{\text{number of correctly recommended items}}{\text{total number of recommended items}} \qquad (5.14)$$

Recall is the number of correct results divided by the number of results that should have been returned and tells us how complete the prediction is.

$$Recall = \frac{\text{number of correctly recommended items}}{\text{total number of relevant items}} \qquad (5.15)$$

By using the *F-measure*, the evaluation results would be more comparable:

$$F_1 = \frac{2 \cdot p \cdot r}{p + r}, \qquad (5.16)$$

where p and r are precision and recall, respectively.

Evaluating the Accuracy of Ranks Precision and recall define the accuracy of prediction. But ranking the scores could be used to evaluate the recommender systems in a finer level of granularity. This lets us consider the position of the recommended item in the final ranking along with the prediction results. We can use different functions to evaluate the accuracy of ranks based on this method. The following equations show an example that decreases the utility value of a recommendation based on the position of the item in the recommended item list [20]:

$$rankScore_u = \sum_{i \in rec_u} (2^{\frac{rank(i)-1}{\alpha}})^{-1} \qquad (5.17)$$

$$rankScore_u^{max} = \sum_{i \in test_u} (2^{\frac{index(i)-1}{\alpha}})^{-1} \qquad (5.18)$$

$$rankScore'_u = \frac{rankScore_u}{rankScore_u^{max}},\qquad(5.19)$$

where rec_u denotes the list of recommended items, $rank(i)$ denotes the position of item i in the list, and $index(i)$ denotes the real position of the item i in the list. The parameter α is a half-life utility which means that an item at position α generates twice the utility as an item at position $\alpha + 1$.

Evaluating the Acceptance Level Evaluating the acceptance level of the recommendation can be more directly measured in industry through the following: *click-through rate, usage, return rate* and *profit*, by checking whether the customer accepted the recommendation, bought the product and did not return the product.

5.1.5 Challenges of Recommender Systems

Though recommender systems have been very successful in past, they have also encountered some challenges. Recommender systems, in particular, collaborative filtering based recommender systems, face three major challenges: *cold start, data sparsity,* and *attacks*.

5.1.5.1 Cold Start

Cold start is a potential problem for any data-driven system, including recommender systems that try to build a model based on the existing information. Cold start is the situation that the algorithm's effectiveness is very low because items' (or users') vectors do not have enough rated items to find vectors similar to them [235]. In the content-based approach, the systems must be able to match items' characteristics against relevant features from users' profiles. Therefore, it needs to construct a detailed model of users' tastes and preferences. Without having a sufficiently detailed model of users' tastes and preferences, the system would fail to match the appropriate items and consequently to make a recommendation for users. In the collaborative filtering approach, the recommender system identifies users who share similar preferences with the active user and recommends items which the like-minded users favored (and the active user has not yet seen). Due to the cold start problem, this approach would fail to consider items which no one in the system has rated previously. The cold start problem can be mitigated by applying hybrid approaches such as a combination of content-based and collaborative filtering approaches.

5.1.5.2 Data Sparsity

The core of many recommender systems is to find similar users or similar items. Though there are many algorithms that can solve this problem, almost all of them fail when the size of the vectors grows and passes some threshold [27]. When the number of users or items increases, the rating matrix becomes extremely sparse. For example, Internet Movie Database (IMDB) has records of more than 700K movies. Even if somebody could see and rank one thousand of them, the rating matrix would remain extremely sparse. In these situations, finding similar users becomes extremely difficult, and most of the existing algorithms fail to find similar users or items. One common technique to handle this problem is using *factorization methods* to reduce the size of the rating matrix and create a matrix with a lower number of more relevant and independent features. However, handling extremely sparse rating matrices remains an open challenge for recommender systems [232].

5.1.5.3　Attacks

Attacks are designed to drive the recommender system to act in a way that the attacker wishes. It could either recommend some desired items or prevent recommending of other items. An attack against a recommender system consists of a set of attack profiles, and each contains biased rating data associated with a fictitious user identity and a target item. There are two major types of attacks: *Push attack* or *Nuke attack* [38, 129]. In a push attack, the recommender recommends whatever the attacker wishes. The attacker creates a profile whose rating is similar to that of the target user. Given the high similarity between the target user and the injected profile, the system will more likely use the injected profile as a source to provide recommendation to the target user. The attacker in a nuke attack aims to prevent the system from recommending a particular item or to make the recommender system unusable as a whole.

5.1.6　Summary

We have discussed two popular types of recommender systems, content-based and collaborative filtering systems. Although these approaches are highly successful and are frequently used in the real-world recommender systems, they have their limitations when handling challenges such as *cold start, data sparsity,* and *attacks.* There are many ongoing research projects to address these challenges. For example, using *trust information* has gained more attention from both academia and industry researchers. Next, we introduce trust in social networks and then approaches of trust-aware recommendation systems.

5.2 Computational Models of Trust in Recommender Systems

Trust plays an important role across many disciplines and forms an important feature of our everyday lives [164]. In addition, trust is a property that associates with the relations between people in the real world as well as users in social media. In the previous section, we describe the challenges of the existing recommender systems including *cold start*, *data sparsity*, and *attacks*. It has been shown that, by using trust information, we can mitigate some of the challenges. In this section, we provide a brief review of trust and solutions which we can use to measure trust among social media users in general and users of recommender systems in particular. This trust information will be used as a base to improve classical recommender systems and construct trust-aware recommender systems, which will be covered in the next section in detail. We start the section by defining trust and its properties. The section also focuses on measuring trust values between users including both explicit and implicit trust values. Propagation and aggregation techniques of trust values are the last important concepts which we cover in this section.

5.2.1 Definition and Properties

Trust, in social context, usually refers to a situation where one party (*trustor*) is willing to rely on the actions of another party (*trustee*). It is also attributed to relationships between people or within and among social groups, such as families, friends, communities, organizations, or companies [143]. In recommender systems, it is defined based on the other users' abilities to provide valuable recommendations [90]. In this work, we adopt the definition given in [172] (Page 712): *"The willingness of a party to be vulnerable to the actions of another party based on the expectation that the other will perform a particular action important to the trustor, irrespective of the ability to monitor or control that other party"*.

In this subsection, we consider only trust that forms among the users of a system, and do not consider other forms of trust such as trust between users and items or users and communities.

5.2.1.1 Notations

In our settings, we use $T(u, v)$ to show there is a trust relation between the truster, u, and the trustee, v, and we use $t_{u,v}$ to indicate the value of the trust between them. The trust value can be either binary or real numbers (e.g., in the range of $[0, 1]$). Binary trust is the simplest way of expressing trust. Either two users trust each other or not. A more complicated method is the continuous trust model, which assigns real values to the trust relations.

In both binary trust and continuous models, 0 and 1 mean *no trust* and *full trust*, respectively. When we consider distrust, we have to add negative values of trust in the range $[-1, 0)$. Therefore, we represent trust (and distrust) as a continuous variable over the range $[-1, +1]$. Many applications, such as Amazon[4] and eBay,[5] use binary trust values. We use $\mathbf{T} \in \mathbb{R}^{n \times n}$ to denote the matrix representation of trust values for a system with n users. In this perspective, trust values are assigned to the relations. Moreover, we can think of trust as a property of users, or nodes in social media. In this situation, every user has a global trust value representing the user's overall trustworthiness. We use t_u to represent the global trust to user u. We use $\mathbf{t} \in \mathbb{R}^n$ to denote the vector representation of global trust values for a system with n users.

5.2.1.2   Trust Networks

A trust network is a weighted and directed graph in which nodes are users and edges are the trust between the users. In a trust network, trust is the information about the social relationships and it is represented as a label for the links. The strength of the edges shows the amount of trust between the users at the two ends of every edge. Figure 5.3 represents a trust network.

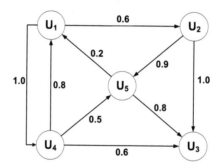

FIGURE 5.3: A trust network.

The matrix representation of the trust network is given in Equation 5.20.

$$\mathbf{T} = \begin{bmatrix} 0 & 0.6 & 0 & 1.0 & 0 \\ 0 & 0 & 0.9 & 0 & 1.0 \\ 0.2 & 0 & 0 & 0 & 0.8 \\ 0.8 & 0 & 0.5 & 0 & 0.6 \\ 0 & 0 & 0 & 0 & 0 \end{bmatrix} \tag{5.20}$$

5.2.1.3   Properties of Trust

In this section, we introduce properties of trust including *transitivity*, *asymmetry*, *context dependence*, and *personalization* [33, 83, 90]. These prop-

[4]www.amazon.com
[5]www.ebay.com

erties are extracted from the definition of trust and provide the basis for the creation of algorithms that utilize trust information.

- **Transitivity** Transitivity is a key property of trust. It allows trust to be propagated along paths to reach other users. Based on the transitivity effect, if u trusts v and v trusts w, it can be inferred that u might also trust w to some extent (see Figure 5.4):

$$T(u,v) \wedge T(v,w) \Rightarrow T(u,w), \qquad (5.21)$$

where $T(u,v)$ indicates trust relation between users u and v, and $t_{u,v} > 0$.

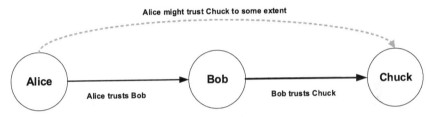

FIGURE 5.4: Transitivity of trust.

- **Composability** Transitivity describes how trust information propagates from one user to another through a chain of connected users. Composability describes how a user should combine the different trust values received from different paths. There are different approaches we can use to aggregate trust values from different paths but the simplest way is to first sum them up and then normalize the values based on the predefined ranges of trust values. Figure 5.5 shows the case that user *Alice* receives trust values from two different paths.

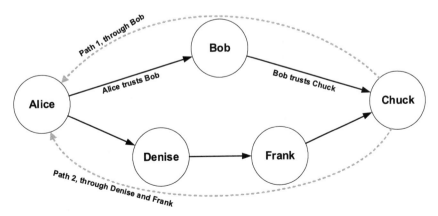

FIGURE 5.5: Composability of trust.

- **Asymmetry** Trust is a subjective and personal relation between users; therefore, it creates a directed relation in social networks. In other words, if $t_{u,v}$ represents the value of trust from user u to user v, it might not necessarily be equal to the value of trust from user v to user u. Based on this attribute, user u trusting user v does not guarantee that user v also trusts user u to the same extent:

$$t_{u,v} \neq t_{v,u}. \tag{5.22}$$

- **Context Dependence** Trust is context dependent. This means that trusting someone on one topic does not guarantee trusting him in other topics. For example, a user who is trustworthy in technology might not also be trustworthy in astronomy, as well:

$$T_i(u, v) \nRightarrow T_j(u, v), \tag{5.23}$$

where $T_i(u, v)$ indicates the existence of trust between users u and v with respect to topic i.

- **Personalization** Trust is a personalized attribute. The amount of trust one person has in another might vary between different people. We use this property to define and formulate local trust. But when we define global trust, which is equivalent to user reputation, we violate this property. In global trust every user has only one trust value that can be used by all other users in the network. We describe global trust values in detail in Section 5.2.2,

$$t_{u,v} \neq t_{w,v}. \tag{5.24}$$

5.2.2   Global and Local Trust Metrics

Based on our earlier definition, trust is a relation between two people and trust values indicate how much one person trusts the another. Trust metrics are the methods for measuring and calculating the value of trust between users in the network. From this point of view, trust is a local attribute [169]. In addition to this local view, we can also look at trust as a global measure, where every user has a global value that indicates his/her trustworthiness in the network as a whole [168]. In this section, we elaborate on these two views of trust.

Local Trust Metrics Local trust metrics use the subjective opinion of the current user to predict the trustworthiness of other users from the current user's perspective. The trust value represents the amount of trust that the active user puts on another user. Based on this approach, different users trust the current user differently, and therefore, their trust values are different from each other $t_{iu} \neq t_{ju}$. Consider a network of n users. In a local trust metric, every user evaluates all other $n-1$ users and assigns them a value that represents their trustworthiness. Therefore the local trust metrics are defined as $T : U \times U \to [0,1]$.

Global Trust Metrics Global trust metrics represent the whole community's opinion regarding the current user; therefore, every user receives only one value that represents her level of trustworthiness in the community. Trust scores in global trust metrics are calculated by the aggregation of all users' opinions regarding the current user. Users' reputations on ebay.com are an example of using global trust in an online shopping Website. ebay.com calculates user reputation based on the number of users who have left positive, negative, or neutral feedback for the items sold by the current user. When the user does not have a specific opinion regarding another user, she usually relies on this aggregated trust scores. The global trust metrics are defined as $T : U \to [0,1]$.

Global trust can be further classified into two classes of *profile-level* and *item-level* trust. The profile-level trust refers to the general definition of global trust metrics in which it assigns one trust score to every user. In item-level trust, every user might have a different trust score for different items. This supports the third property of trust, which is context dependence.

Comparison between Local and Global Trust Metrics Global metrics are computationally less expensive and simpler than local metrics. However, local models can handle controversial topics and resistance against attacks. In some cases, such as controversial topics, it is difficult to reach an agreement among the users regarding one specific user. In these cases, there are groups of people who like and trust one specific user, and many others dislike or distrust the user. These are situations in which it is highly recommended to use local trust metrics, if applicable. In a local setting, trust propagates only through users who are explicitly being trusted by the current user. Therefore, fake profiles cannot exert their influence on the final propagation and trust scores of other users. As a result, compared to global trust metrics, local trust metrics are shown to be more resistant against the attacks [85,169]. Calculating local trust is computationally more expensive than global trust scores, as it should be calculated for every pair of users. However, it might represent a user's interests more accurately and precisely than the global approach. It is notable that a majority of trust metrics proposed in literature are global and only a few have used local metrics.

5.2.3 Inferring Trust Values

In a network of users, it is possible to ask users to evaluate other users and assign each user a value that indicates his trustworthiness. Although this setting is feasible in the real world, every user can evaluate only a few other users. This might be either due to users' time limitations or because a user

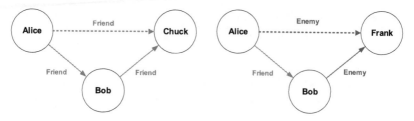

FIGURE 5.6: Trust inference.

has a direct opinion about only a few other users, e.g., on epinions.com, the average size of trusted users for every node is 1.7 [3, 83, 257]. To expand the coverage of users' trust network, computational tools are designed to predict the trustworthiness of unknown users. These methods aim to predict the trust scores for almost every pair of the nodes in the network based on the existing trust information. By using trust inference algorithms we can accurately calculate how much a person should trust another person in the network.

We are interested to know if one can use available trust information to infer trust values between other pairs of the nodes in the network. Given a network of users [197], transitivity is the common assumption in all of these solutions for propagating and infering trust in the network. Based on transitivity, if user u trusts user v and user v trusts user w, we can conclude that user u might also trust user w to some extent. Trust inference techniques can be used to overcome the sparsity and cold start problems in the network.

The following are examples of how we use transitivity in everyday conversation:

- My friend's friend is my friend (left graph in Figure 5.6).

- My enemy's enemy is my friend (right graph Figure 5.6).

In the following section we introduce algorithms to infer trust between users in social networks. All of the algorithms use paths in the network to propagate and infer trust values. We introduce solutions to infer binary trust values as well as continuous trust values.

5.2.3.1 Inferring Trust in Binary Trust Networks

In a binary trust network, nodes are rated as either "trusted" or "not trusted". In this section, we present a simple voting algorithm to infer binary trust values. In Figure 5.7, the goal is to infer $t_{s,t}$, where s denotes the source, t denotes the target node, and $t_{s,t}$ denotes the trust value from s to t. In this algorithm, the source asks all of its "trusted" neighbors to return their ratings for the target node. The source does not ask "not trusted" neighbors, as their ratings would be unreliable for him. If the neighbor is connected to the target node, it simply returns its own trust value (either trusted or not trusted). Otherwise, it repeats the same process by asking neighboring nodes.

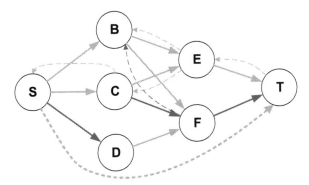

FIGURE 5.7: Inferring trust in a binary trust network.

When a node receives more than one rating from neighbors, it averages the ratings and passes the result to the requester. In the next step, the node can either round the final ratings and pass binary trust values (as $\{0, 1\}$) or pass the real number value (in range $[0, 1]$). Finally, the source takes the average of the values received from different paths and rounds them. The final rating value is either 1 or 0, representing "trusted" or "not trusted", respectively.

In the graph illustrated in Figure 5.7, user S asks B and C for recommendations about T, but ignores D as he does not trust D. Neither B nor C is directly connected to T; therefore, they ask E and F for recommendations. Following the algorithm described above, B receives one "positive" and one "negative" recommendations and C receives one "positive" recommendation. The final result is one positive recommendation from C to S. Therefore, S's evaluation for T is "Trust".

5.2.3.2 Inferring Trust in Continuous Trust Networks

In a network with continuous trust values, taking a simple average of all trust values is the most intuitive approach to infer trust values. In this approach, the source node polls each of its neighbors to obtain their ratings of the target. Each neighbor repeats the same process. Every time a node receives ratings from more than one neighbor, it uses the average value of the ratings. In this approach, we do not consider the length of paths and weights of each neighbor, but treat them all equally. This approach is usually used as a baseline for other solutions on inferring trust in continues trust networks:

$$
t_{u,v} = \frac{\displaystyle\sum_{i \,\in\, adj(u)} t_{u,i} t_{i,v}}{\displaystyle\sum_{i \,\in\, adj(u)} t_{u,i}}, \tag{5.25}
$$

where $adj(u)$ denotes a set of users adjacent to u and $t_{u,i}$ denotes the trust between users u and v.

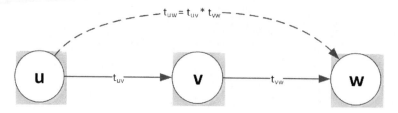

FIGURE 5.8: Trust propagation in a path.

Trust propagation in a path [277] When two users are connected through other users located in a path, we can use propagation models to calculate trust between them. For example, in Figure 5.8 there is no direct connection between users u and w, however, they are connected through user v. We can compute the trust values along the path u, v, w as follows:

$$t_{u,w} = t_{u,v} \times t_{v,w}. \tag{5.26}$$

In general, for a path of length n, trust can be computed as follows:

$$t_{u,w} = \prod_{(a_k,a_l) \,\in\, \text{path } (u,w)} t_{a_k,a_l}, \tag{5.27}$$

where $t_{u,v}$ is the trust value between users u and v and $0 \leq t_{u,v} \leq 1$. Note that we cannot use this technique to propagate distrust. In addition to the simple average approach, we introduce some representative approaches for computing indirect trust as follows: *EigenTrust, Trust Propagation, Appleseed* and *Matrix Factorization-based method*.

EigenTrust The EigenTrust algorithm is a reputation management algorithm for social networks. The algorithm provides each user in the network a unique global trust value based on the user's history of interaction among the users [119]. This approach relies on computing the principal eigenvector of the trust network to calculate trust values. The result of the algorithm is a vector of trust values, representing how much trust the source node should have for other nodes in the network. The values are ranks of the trustworthiness of individuals, not recommended trust values. The biggest challenge with this algorithm is that it needs to convert these ranks into trust values that can be assigned to the relations.

Trust Propagation In Guha et al. [87], a trust propagation framework is proposed with four atomic propagations: direct propagation, co-citation, transpose trust and trust coupling:

- if u_i trusts u_j, and u_j trusts u_k, direct propagation allows us to infer that u_i trusts u_k, and its corresponding operator is **T**.

- co-citation propagation concludes that u_ℓ should trust u_j if u_i trusts u_j and u_k, and u_ℓ trusts u_k. $\mathbf{T}^\top \mathbf{T}$ denotes the operator of co-citation propagation.

- in transpose trust, u_i's trust of u_j causes u_j to develop some level of trust towards u_i. The operator is shown by $\mathbf{T}^\top$.

- trust coupling suggests that u_i and u_j trust u_k, so trusting u_i should imply trusting u_j. $\mathbf{TT}^\top$ denotes the operator.

$\mathbf{C}$ is defined as a single combined matrix of all four atomic propagations,

$$\mathbf{C} = \alpha_1 \mathbf{T} + \alpha_2 \mathbf{T}^\top \mathbf{T} + \alpha_3 \mathbf{T}^\top + \alpha_4 \mathbf{TT}^\top, \tag{5.28}$$

where α_1, α_2, α_3 and α_4 control contributions from direct propagation, co-citation, transpose trust and trust coupling, respectively.

Let $\mathbf{C}^k$ be a matrix where $\mathbf{C}_{ij}^k$ denotes the propagation from u_i to u_j after k atomic propagations. Then, the final estimated matrix representation of the user–user trust relation $\tilde{\mathbf{T}}$ is given by [87]

$$\tilde{\mathbf{T}} = \sum_{k=1}^{K} \gamma^k \mathbf{C}^k, \tag{5.29}$$

where K denotes the number of steps of propagation and γ^k denotes a discount factor to penalize lengthy propagation steps.

Appleseed The Appleseed algorithm proposed by Ziegler in 2004 [301]. It models trust as energy and injects it from the source node into all of the other nodes in the network along the edges. The energy is fully divided among the successor nodes with respect to their local trust scores. In the Appleseed algorithm, the closer the sink is to the source and the more paths leading from the source to the node, the higher the energy that reaches to the sink.

Matrix factorization-based method [257] A few factors can influence the establishment of trust relations. A user usually establishes trust relations with a small proportion of $\mathcal{U}$, resulting in very sparse and low-rank $\mathbf{T}$; hence, users can have a more compact but accurate representation in a low-rank space [262]. The matrix factorization model seeks a low-rank representation $\mathbf{U} \in \mathbb{R}^{n \times d}$ with $d \ll n$ for $\mathcal{U}$ via solving the following optimization problem:

$$\min_{\mathbf{U}, \mathbf{V}} \quad \|\mathbf{T} - \mathbf{UVU}^\top\|_F^2, \tag{5.30}$$

where $\| \cdot \|_F$ denotes the Frobenius norm of a matrix and $\mathbf{V} \in \mathbb{R}^{d \times d}$ captures the correlations among their low-rank representations such as $\mathbf{T}_{ij} = \mathbf{U}_i \mathbf{VU}_j^\top$. It is easy to verify that Equation (5.30) can model the properties of trust such as transitivity and asymmetry [262].

To avoid overfitting, we add two smoothness regularization components on $\mathbf{U}$ and $\mathbf{V}$ in Equation (5.30),

$$\min_{\mathbf{U},\mathbf{V}} \quad \|\mathbf{T} - \mathbf{U}\mathbf{V}\mathbf{U}^{\top}\|_F^2 + \alpha\|\mathbf{U}\|_F^2 + \beta\|\mathbf{V}\|_F^2, \qquad (5.31)$$

where α and β are nonnegative and are introduced to control the capability of $\mathbf{U}$ and $\mathbf{V}$, respectively. With the learned $\mathbf{U}$ and $\mathbf{V}$, the estimated matrix representation of the user-user trust relation $\tilde{\mathbf{T}}$ is obtained as $\tilde{\mathbf{T}} = \mathbf{U}\mathbf{V}\mathbf{U}^{\top}$.

5.2.3.3 Inferring Implicit Trust Values

In implicit trust models, there is no trust information in the system, and the goal is to use other information to construct a network with trust values between the users. Ziegler [302] showed that there is a relationship between people's interest similarities and trust between them. Jensen, Davis and Farnham [108] found that given some predefined domain and context, people's interest similarity is a strong predictor of interpersonal trust. This means that people who have similar interests tend to be more trustful towards each other. In the light of these studies, it can be concluded that measuring the similarity in interests is a reasonable technique to measure trust among the users.

The models use different network information to calculate the trust between users. In this section we introduce the techniques of *profile similarity*, *rating history*, and *relevance* which have received significant attention in the literature.

- **Profile similarity** — Similarity between users (network, content, or interests). If two people are connected, consider this as a sign for their trust. This way trust can be mutual or directed.

- **Rating history** — Calculating the similarity based on the similarity in product ratings. Users with similar ratings and similar tastes are more likely to trust each other.

- **Relevance** — Using the recommendation history to measure relevance based on the accuracy of recommendations. If user u has a record of successful recommendations to user v, this can be used as a sign that v trusts u.

There is a major difference between implicit trust and inferred trust values. In inferring explicit trust, we use transitivity to propagate existing trust information and calculate the trust information for users with missing trust information. Unlike the explicit inference method, in implicit trust inference, we do not use any other trust information in the network.

5.2.3.4 Trust Aggregation

Another aspect of trust inference is aggregating trust values inferred from other nodes. When a user asks for a recommendation and more than one neighbor answers or there is more than one recommendation per neighbor, we need to integrate the recommendations to end up with one value. To make the final choice, we can use any of the following approaches [170]:

- Average to get the average of all of the recommendations,

- Frequency-based approach to get the trust value with the highest frequency,

- Shortest path to rank recommendations based on the length of the path to the target user (i.e., shorter paths will get higher weight),

- Top-Friends suggestion to give higher scores to the values recommended by closer friends, and

- Hybrid to use a combination of the above approaches.

5.2.4 Summary

In this section, we have discussed trust in social networks. We have formally defined trust and introduce its properties. We have provided an overview on measuring explicit and implicit trust in the network and discussed algorithms for inferring trust in explicit networks. In the next section, we introduce techniques of using trust formation to improve recommender systems.

5.3 Incorporating Trust in Recommender Systems

Users can be influenced by their trustworthy friends and are more likely to accept recommendations made by their trusted friends than recommendations from strangers. Studies also confirm that people tend to rely on recommendations from their friends and other people they trust more than those provided by recommender systems in terms of quality and usefulness, even though the recommendations given by the recommender systems have high novelty factor [246]. *Trust-aware recommender systems* employ trust information to enhance recommender systems. Merging trust information and recommender systems can improve the accuracy of recommendations as well as the users' experiences.

In addition to these improvements, which will be discussed in this section, trust-aware recommender systems are capable of handling some of the challenges we encounter for classical recommender systems such as the cold start problem and responding to attacks. Trust metrics and recommendation technologies are two bases for trust-aware recommender systems [273] (see Figure 5.9). The focus of the section is describing techniques to combine trust in-

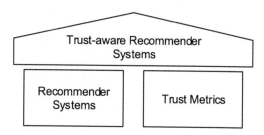

FIGURE 5.9: A trust-aware recommender system.

formation with classical recommender systems. As mentioned earlier, most of the trust-aware recommender systems are based on the collaborative filtering approach.

Trust-aware recommender systems utilize trust information to generate more personalized recommendations, and users receive recommendations from those who are in their trust networks (Web of trust). Trust-aware recommender systems emerge from social networks where there is the information about the ties between users. This information can be used to construct a user–user similarity bases. This extra information will be added to traditional recommender systems to improve the quality of the recommendation, in hopes of improving the *accuracy* or *coverage* and addressing some of the challenges from the classical recommender systems. In a trust-aware recommender system, trust information can be used in one of the following approaches along with classical recommender systems:

- *Trust-aware memory-based CF systems:* Systems of this group use memory-based CF techniques as their basic methods.

- *Trust-aware model-based CF systems:* Systems of this group use model-based CF techniques as their basic methods.

FilmTrust and *epinions.com* are two examples of trust-aware recommender systems.

FilmTrust [81] is an example of a trust-aware recommender system (see Figure 5.10). The system is an online social network that combines trust, social networks, and movie ratings and reviews to generate a personalized recommendation. In FilmTrust, personalization of the Website, displaying movie ratings, and ordering the reviews are based on the trust and rating information. There is a social network and for each connection, users rate how much

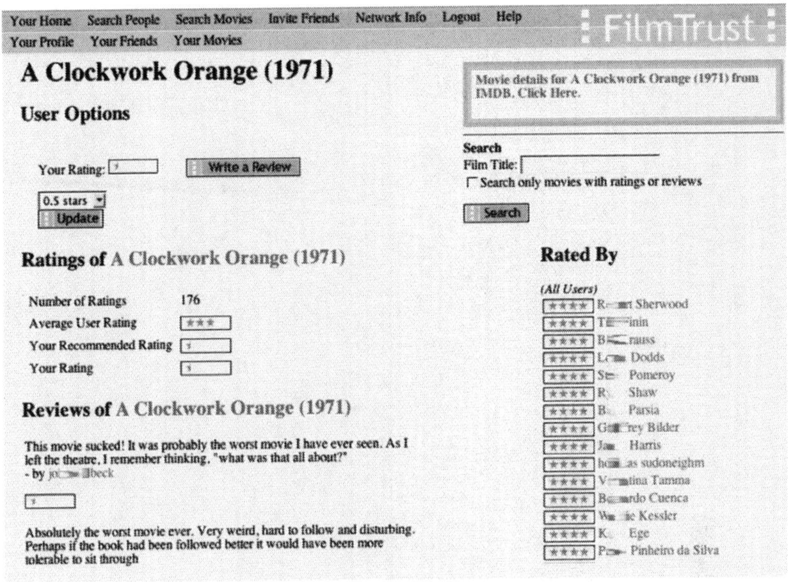

FIGURE 5.10: FilmTrust.

they trust their friend's opinion about movies. The following sentence from the help section of the systems would help users on rating their friends: *"Think of this as if the person were to have rented a movie to watch, how likely it is that you would want to see that film."* Users can evaluate their friends' movie tastes on a scale from 1 (low trust) to 10 (high trust). But these numbers remain private and other users cannot see the actual values.

Users also rate movies and write reviews as well. These ratings are on a scale of a half star to four stars. When a user asks for a recommendation, it shows two sets of information for every movie, the overall average rating and the recommended rating calculated using trust values as weights. FilmTrust uses the trust information to weight the ratings of users in the network (similar to trust-based weighted mean).

www.epinions.com[6] is a general consumer review site (see Figure 5.11). Users in this product review site can share their reviews about products. Also they can establish their trust networks from which they may seek advice to make decisions. People not only write critical reviews for various products but also read and rate the reviews written by others. Furthermore, people can add members to their trust networks or "Circle of Trust", if they find their reviews consistently interesting and helpful.

---

[6]http://www.epinions.com

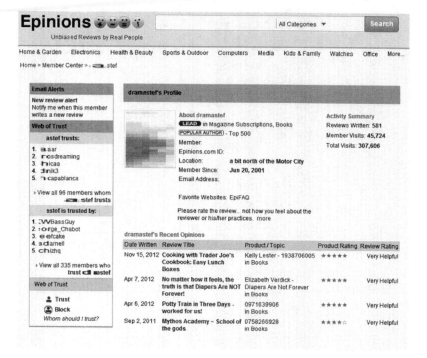

FIGURE 5.11: Epinions, an example of a trust-aware recommender system.

5.3.1 Trust-Aware Memory-Based CF Systems

A different approach from traditional memory-based collaborative filtering is to focus on friends more than strangers. In this approach, the system incorporates trust information to boost recommendations from trusted users and depress recommendations from other users. This recommender system uses trust information to either weight the recommendation made be all users or to filter nontrusted users.

5.3.1.1 Trust-Aware Filtering

Pure collaborative approaches use all available data (users or items) to make recommendations regardless of the importance of each of them to the receiver of the recommender. In trust-aware filtering, we use trust information as a tool to filter some of the users, then we use recommendations provided by trusted users. After filtering the extra users, we use classical recommendations approaches to recommend items. Therefore, the trust-aware filtering model accepts only the recommendations made by trusted users. This approach is more effective when the binary values of trust are available. In the case that we have continues trust values, the system filters users that the current user

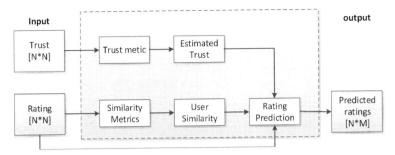

FIGURE 5.12: Trust-aware weighting.

does not trust or for whom the trust value is less than a threshold as follows:

$$\hat{r}_{u,i} = \bar{r}_u + \frac{\sum_{t_{u,v} > \tau} (r_{v,i} - \bar{r}_v)}{\sum_{t_{u,v} > \tau} r_{v,i}}, \tag{5.32}$$

where $\bar{r}_u$ denotes the average rating for user u.

5.3.1.2 Trust-Aware Weighting

Despite the trust-aware filtering that excludes nontrusted users, trust-aware weighting uses trust information to weight the recommendations made by all users. In this approach, the system incorporates trust information to boost recommendations from trusted users and depress recommendations from nontrusted users. Among the trusted users, those with higher trust values will have more weight on recommending new items to the user. In trust-aware weighting approach, the recommended ratings $\hat{r}$ for user u and item i is calculated using the following equation:

$$\hat{r}_{u,i} = \bar{r}_u + \frac{\sum (r_{v,i} - \bar{r}_v) \times t_{u,v}}{\sum t_{u,v}}, \tag{5.33}$$

where $\hat{r}_{u,i}$ denotes user u's predicted rating for item i, and $t_{u,v}$ denotes the trust value from user u to user v. Another approach is to use a combination of filtering and weighting approaches as follows:

$$\hat{r}_{u,i} = \bar{r}_u + \frac{\sum_{t_{u,v} > \tau} (r_{v,i} - \bar{r}_v) \times t_{u,v}}{\sum_{t_{u,v} > \tau} t_{u,v}}. \tag{5.34}$$

The rest of this section introduces some of the representative trust-aware memory-based recommender systems.

TidalTrust TidalTrust [83] computes trust scores by performing a modified breadth first search in a trust network (see Figure 5.13). In this algorithm, the goal is to compute the trust values between a source and a sink node,

$t_{source,sink}$. The algorithm starts with the source node and finds all the paths to the sink. Every path has two properties: its length and its strength. Strength shows the trust score of the source and the current node which is equal to the minimum trust scores of all edges located on the path. Among the paths, those with the shortest path length will be selected. This algorithm performs in two steps. First, it finds all raters with the shortest path distance from the source user to the sink. Then, it aggregates their ratings weighted by the trust between the user and the raters. The following two principles are features integrated into TidalTrust:

- For a fixed trust value, shorter paths have lower error.

- For a fixed path length, higher trusts have lower error.

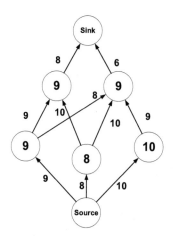

FIGURE 5.13: TidalTrust.

MoleTrust MoleTrust is a local trust metric proposed by Massa and Avesani [169]. The idea of MoleTrust is similar to TidalTrust, except that the depth of the search path for rater can be up to a *maximum-depth*. To calculate the trust scores, the algorithm performs a backward exploration:

$$t_u = \frac{\sum_{i \in predecessors} (t_i \times t_{i,u})}{\sum_{i \in predecessors} (t_i)}, \tag{5.35}$$

where t_i denotes user i's global trust value. In MoleTrust and TidalTrust, if a node does not know the sink, it asks its friends to learn how much to trust the sink. Then, it computes a trust value by a weighted average. If neighbors are directly connected to the sink, they return their ratings, otherwise they repeat same process by asking their own neighbors:

$$t_{i,s} = \frac{\displaystyle\sum_{j \,\in\, adj(i) \;|\; t_{i,j} \,\geq\, max} t_{i,j}t_{j,s}}{\displaystyle\sum_{j \,\in\, adj(i) \;|\; t_{i,j} \,\geq\, max} t_{i,j}}. \tag{5.36}$$

TrustWalker [104] The intuition of this system is from two key observations. First, a user's social network has little overlap with similar users, suggesting that social information provides an independent source of information. Second, ratings from strongly trusted friends on similar items are more reliable than ratings from weakly trusted neighbors on the same target item. The first observation indicates the importance of trust-based approaches, while the second observation suggests the capability of item-oriented approaches. To take advantage of both approaches, TrustWalker proposes a random walk model to combine trust-based and user-oriented approaches into a coherent framework. It queries a user's direct and indirect friends' ratings for the target item as well as similar items by performing a random walk in online social networks. For example, to obtain a rating for u_i to v_j, suppose that we are at a certain node u_k, then TrustWalker works as follows at each step of a random work: if u_k rated v_j, then it stops the random walk and returns $\mathbf{R}_{kj}$ as the result of random walk; otherwise, it has two choices: (1) it also stops the random walk and randomly selects one of the items v_k similar to v_j rated by u_i and returns $\mathbf{R}_{ik}$, or (2) it continues random walk and walks to another user u_k in u_i's trust network. TrustWalker employs the Pearson Correlation Coefficients of ratings expressed for items to calculate item–item similarity. Since values of Pearson Correlation Coefficients are in the range of $[-1, 1]$, only items with positive correlation with the target item are considered. The similarity between v_i and v_j is then computed as follows:

$$sim(i, j) = \frac{1}{1 + e^{-\frac{N_{ij}}{2}}} \times PCC(i, j), \tag{5.37}$$

where N_{ij} denotes the number of users who rated both v_i and v_j, and $PCC(i, j)$ denotes the Pearson Correlation Coefficient of v_i and v_j.

5.3.2 Trust-Aware Model-Based CF Systems

Model-based social recommender systems choose model-based CF techniques as their basic models. Matrix factorization techniques are widely used in model-based CF methods. The common rationale behind these solutions is that users' preferences are similar to or influenced by their trusted users. Systems in this class can be further divided into three groups [261]: co-factorization methods, ensemble methods, and regularization methods. Next, we review some representative systems in detail for each group.

Co-factorization methods [159, 257] The underlying assumption of systems in this group is that the i-th user u_i should share the same user preference vector $\mathbf{u}_i$ in the rating space (rating information) and the trust relation space. Systems in this group perform a co-factorization in the user–item matrix and the user–user trust relation matrix by sharing the same user preference latent factor. For example, the representative system SoRec [159] learns the user preference matrix $\mathbf{U}$ from both rating information and trust information by solving the following optimization problem:

$$\min_{\mathbf{U},\mathbf{V},\mathbf{Z}} \|\mathbf{W} \odot (\mathbf{R} - \mathbf{U}^\top \mathbf{V})\|_F^2 + \alpha \sum_{i=1}^{n} \sum_{u_k \in \mathcal{N}_i} (\mathbf{T}_{ik} - \mathbf{u}_i^\top \mathbf{z}_k)^2$$
$$+ \lambda(\|\mathbf{U}\|_F^2 + \|\mathbf{V}\|_F^2 + \|\mathbf{Z}\|_F^2). \tag{5.38}$$

The reconstructed matrix $\hat{\mathbf{T}}$ can be used to perform trust relation prediction [176, 257]. Therefore, one advantage of approaches in this group is that they jointly perform recommendation and trust relation prediction.

Ensemble methods [157, 258] The basic idea of ensemble methods is that users and their trust networks should have similar ratings on items, and a missing rating for a given user is predicted as a linear combination of ratings from the user and her trust network. *STE* [157] is a representative system in this group, which models a rating from the ith user u_i to the jth item v_j as follows:

$$\hat{\mathbf{R}}_{ij} = \mathbf{u}_i^\top \mathbf{v}_j + \beta \sum_{u_k \in \mathcal{N}_i} \mathbf{T}_{ik} \mathbf{u}_k^\top \mathbf{v}_j, \tag{5.39}$$

where $\sum_{u_k \in \mathcal{N}_i} \mathbf{T}_{ik} \mathbf{u}_k^\top \mathbf{v}_j$ is a weighted sum of the predicted ratings for v_j from u_i's trust network, and β controls the influence from trust information. STE finally minimizes the following term:

$$\|\mathbf{W} \odot ((\mathbf{R} - \mathbf{U}^\top \mathbf{V}) - \beta \mathbf{T}\mathbf{U}^\top \mathbf{V})\|_F^2 \tag{5.40}$$

Regularization methods [105, 160] Regularization methods focus on a user's preferences and assume that a user's preferences should be similar to that of her trust network. For a given user u_i, regularization methods force her preferences $\mathbf{u}_i$ to be closer to those of users in u_i's trust network $\mathcal{N}_i$. SocialMF [105] forces the preferences of a user to be closer to the average preference of the user's trust network as

$$\min \sum_{i=1}^{n} (\mathbf{u}_i - \sum_{u_k \in \mathcal{N}_i} \mathbf{T}_{ik} \mathbf{u}_k)^2, \tag{5.41}$$

where $\sum_{u_k \in \mathcal{N}_i} \mathbf{T}_{ik} \mathbf{u}_k$ is the weighted average preference of users in u_i's trust network $\mathcal{N}_i$. The authors demonstrated that SocialMF addresses the transitivity of trust in trust networks because a user's latent feature vector is dependent on the direct neighbors' latent feature vectors which can propagate in the entire network. In this case, the user's latent feature vector is dependent on possibly all users in the network. SocialMF solves the following optimization problem:

$$\min_{\mathbf{U},\mathbf{V}} \|\mathbf{W} \odot (\mathbf{R} - \mathbf{U}^{\top}\mathbf{V})\|_F^2 + \alpha \sum_{i=1}^{n} (\mathbf{u}_i - \sum_{u_k \in \mathcal{N}_i} \mathbf{T}_{ik}\mathbf{u}_k)^2$$
$$+ \lambda(\|\mathbf{U}\|_F^2 + \|\mathbf{V}\|_F^2) \tag{5.42}$$

One advantage of these approaches is that they indirectly model the propagation of tastes in social networks, which can be used to mitigate cold-start problem and increase the coverage of items for recommendations.

5.3.3 Recommendation Using Distrust Information

Distrust information considers disagreements between users. While trust information shows to what degree one user likes recommendations made by another user, distrust information shows the opposite. By considering distrust information, recommender systems can avoid getting recommendations from users that the current user distrusts. Distrust information also helps recommender systems learn about the items that the user does not like. This comes from the idea that if a user does not trust another user, she does not like items recommended by the user. So the recommender system will avoid recommending those items as well. There are two strategies to exploit distrust information for recommendation including *distrust as filter* and *distrust as an indicator to reverse deviations* [271].

Distrust as a filter: This strategy is to use distrust to filter out "unwanted" individuals from collaborative recommendation processes as follows:

$$\hat{r}_{u,i} = \bar{r}_u + \frac{\sum_{v \backslash \mathcal{D}} (r_{v,i} - \bar{r}_v) \times t_{u,v}}{\sum t_{u,v}}, \tag{5.43}$$

where $\mathcal{D}$ is the set of distrusted users.

Distrust as an indicator to reverse deviations: This strategy directly incorporates distrust into the recommendation process by considering distrust scores as negative weights as follows:

$$\hat{r}_{u,i} = \bar{r}_u + \frac{\sum_{v \backslash \mathcal{D}} (r_{v,i} - \bar{r}_v) \times t_{u,v}}{\sum t_{u,v}} - \frac{\sum_{v \in \mathcal{D}} (r_{v,i} - \bar{r}_v) \times d_{u,v}}{\sum d_{u,v}}, \tag{5.44}$$

where $d_{u,v}$ is the distrust strength between u and v.

5.3.4   Advantages of Trust-Aware Recommendation

The main idea behind trust-aware recommendation is to use trust information to improve the quality of recommendation and to address some of the challenges from classical recommender systems especially the cold-start problem and their weakness against attacks. In this section, we introduce some of the advantages of using trust-aware recommender systems.

- *Mitigating the cold-start problem.* New users have only a few or even no ratings. The classical recommender systems may fail to do recommendations for these users. However, trust-aware recommender systems can make recommendations as long as a new user establishes trust relations with other users. Massa and Avesani [166] showed that trust-aware recommendation yields more accurate predictions for cold-start users, compared to a classical collaborative filtering system.

- *Robust to attacks.* By using trust information, we can filter most fake profiles, as no one willfully trusts a fake profile. In this case, the system will not use these users in recommendation.

- *Increasing the coverage of recommendation.* In classical collaborative filtering, we have access only to immediate neighbors who have similarly rated a product. But trust-aware recommender systems use more information to construct the similarity matrix. Therefore, we have access to more users [104]. In addition, trust information increases the coverage of a collaborative filtering system, while maintaining the accuracy [167].

5.3.5   Research Directions of Trust-Aware Recommendation

In this section, we discuss several research directions that can potentially improve the capabilities of trust-aware recommender systems and make social recommendations applicable to an even broader range of applications.

Temporal Information Customer preferences for products drift over time. For example, people interested in "Electronics" at time t may shift their preferences to "Sports" at time $t + 1$. Temporal information is an important factor in recommender systems. Some of the traditional recommender systems already considered temporal information [56, 123]. Temporal dynamics in data can have a more significant impact on accuracy than designing more complex learning algorithms [123]. Exploiting temporal information for recommender systems is still challenging due to the complexity of users' temporal patterns [123].

Trust relations also vary over time. For example, there might be a case where new trust relations are added, while existing trust relations become inactive or are deleted. The variation of both ratings and trust relations further exacerbate the difficulty of exploiting temporal information for social recommendation. A preliminary study of the impact of the changes in both ratings and trust relations on recommender systems demonstrates that temporal information can benefit trust recommendation [259].

Distrust in Recommendation Currently most existing trust-aware recommender systems use trust relations. However, distrust in recommendation is still in the early stages of development and an active area of exploration. Abbassi, Aperjis and Huberman in [4] found that negative relations such as distrust are even more important than positive relations, revealing the importance of distrust relations for recommendation. There are several works exploiting distrust [158, 272] in social recommender systems. They treat trust and distrust separately, and simply use distrust in an opposite way to trust, such as filtering distrusted users or considering distrust relations as negative weights. However, trust and distrust are shaped by different dimensions of trustworthiness. Further, trust affects behavior intentions differently from distrust [39]. Furthermore, distrust relations are not independent of trust relations [273]. A deeper understanding of distrust and its relations with trust can help to develop efficient trust-aware recommender systems by exploiting both trust and distrust.

5.4 Conclusion

Recommender systems aim to overcome the information overload problem in a way that covers users' preferences and interests. Trust-aware recommender systems are an attempt to incorporate trust information into classical recommender systems to improve their recommendations and address some of their challenges including the cold-start problem and their weakness against the attacks.

Chapter 6

Biases in Trust-Based Systems

6.1   Introduction

With the increasing prevalence of social information systems, users increasingly turn to the Web for information generated by other fellow users. This includes looking up online reviews (e.g., Amazon.com) in deciding which product to purchase, bookmarking sites (e.g., Delicious.com) in deciding which Webpage to read, location-based social networks (e.g., Foursquare.com) in deciding which point of interest to visit, etc.

Because all this information come from other users, its utility depends on how much we trust the sources of information. Trust plays a significant role in helping users to filter the deluge of information. Evidently, an online review from a trusted contributor may sway our opinion, while other reviews from those whom we do not trust may not have much of an effect.

Key to the notion of trust is our perception of the honesty and truthfulness of the information sources, that they are *unbiased*. We would like to know whether the information sources have a truly objective view, not influenced by prejudice or other vested interests. However, "bias" is insidious. It is difficult to know when an information source is unbiased. In some cases, the information source itself may not be cognizant of the potential biases that she has. Therefore, identifying occurrences of biases and addressing them effectively, are useful to ensure that as much as possible we obtain unbiased feedback when consulting social information systems.

To place our discussion on a concrete setting, let us consider an online rating system, with a set of reviewers, each denoted r_i, and a set of objects, each denoted o_j. In this rating system, each reviewer r_i may assign a rating score $e_{ij} \in [0,1]$ to an object o_j, which evaluates the quality of o_j in the assessment of r_i. In addition to the rating score, r_i may also write a text review that accompanies the rating. The goal is to arrive at an "unbiased" estimation of the quality q_j of o_j based on the collective assessments of various reviewers.

This particular setting for studying biases is applicable to many scenarios, such as online reviews and academic peer reviews. In addition to deriving an unbiased estimation of object quality, this study may also have implications to tasks such as assigning the appropriate reviewers to objects to be evaluated (e.g., grant applications [60, 98]). A better estimation of object quality may also contribute to more accurate assessment of user preferences in recommender systems [220] (since ratings also indicate subjective preference, in addition to objective evaluation).

The rest of this chapter will be organized as follows. Section 6.2 identifies different types of biases that may affect a rating system. In particular, a distinction is drawn based on intent, distinguishing cognitive biases (non-malicious) from spam (malicious). Section 6.3 discusses the question of how to detect occurrences of biases on the part of reviewers, outlining both unsupervised and supervised approaches that appear in the literature for bias detection. Section 6.4 describes ways to lessen the impact of biases, if they cannot be removed altogether. Finally, Section 6.5 concludes with a summary.

6.2   Types of Biases

To describe different types of biases, it is helpful to differentiate between the cases where biases have malicious intent, referred to as *spam*, from the cases where biases are due to human tendencies when thinking, referred to as *cognitive biases*.

6.2.1   Cognitive Bias

Cognitive biases are various effects that influence the thinking of human subjects in decision making. This is an active field of study in psychology and behavioral economics [269]. There are many types of such biases, and new ones are still being identified. The following is an (incomplete) list of these biases, especially those that affect trust-based systems such as online ratings and reviews.

- *Simple bias* is the tendency to systematically overestimate or underestimate the quality of one's experience [244]. In a rating system, this may manifest as consistently giving ratings that are too high or ratings that are too low. Taking the ratings literally may result in a misreading of the quality of the objects being evaluated.

- *Sequential bias* is the tendency to base one's rating of an object on the previous ratings of the same object [210, 245]. In some cases, this may be a manifestation of the "bandwagon effect", or group think. One feels unease at expressing a rating that substantially differs from the majority of ratings that have been assigned to the object.

- *Dependency bias* is the tendency to rely too heavily on some factors when making decisions. An example of this is the study of rating dependencies [40, 137], which attempt to discover the extent to which different factors may contribute to the overall rating. For example, a frugal user's rating on a restaurant may be overly influenced by the price. In some cases, this may be a manifestation of the "anchoring" effect, in which one anchors one's decision on certain factors. In other cases, this could be a manifestation of the "halo" effect. For instance, someone who loves the brand *Sony* may be positive about all things made by Sony, regardless of the quality of individual products.

- *Underreporting bias* happens when a subset of users choose not to present any rating or evaluation [244]. For example, while many people have been to hotels, in relative terms, very few people have ever reviewed or rated hotels on a Website such as TripAdvisor. People who are most motivated to create reviews could be those with extremely positive or extremely negative experiences that move them to take the trouble to evaluate. This may result in a distribution of ratings or reviews on the Website that is different from the true distribution.

- *Sample size bias* occurs when one is not aware that variance is much more likely in small samples [268]. For example, suppose that a collection contains an even (50:50) number of high-quality and low-quality items. If every reviewer is randomly assigned only a small sample of this collection, it is likely that many reviewers will not get an even split between the two quality levels, but will rather get a skewed distribution. Some may see a skewed number of low-quality items. Others may see a skewed number of high-quality items. These variances in turn affect the perception of individual reviewers.

- *Comparative bias* occurs when one is using different points of comparison when assigning a rating. For example, the "decoy effect" [213] documents the case where if one cannot decide between two items A and B, then being presented with an alternative C (that is clearly worse than B) may sway one's opinion in favor of B. This implies that our evaluation of things is heavily influenced by the set of alternatives that we face.

The above biases are cognitive in nature. In other words, they occur naturally to human subjects in terms of their thinking processes. The implication is that they are not *malicious* in intent.

6.2.2   Spam

In addition to the nonmalicious cognitive biases, there are other types of "biases" that seek to mislead other users deliberately. These are usually referred to as *spam*. They are prevalent in some trust systems such as ratings and reviews, because most Web users rely heavily on them in their decision making. These spams may look similar to cognitive biases, and it is difficult to infer the intent behind a rating or a review.

Nevertheless, several types of spams affecting ratings and reviews have been documented, as follows.

- *Disruptive spam* refers to certain acts by spammers that disrupt the experience of review readers [109]. This could for instance be in the form of reviews only about the brand, but not about the product. Another instance is posting duplicate reviews across multiple products. While given the right context, reviewer readers may be able to detect such spam, they still disrupt the experience of review readers.

- *Distributional spam* occurs when spammers try to influence the rating distribution of a given product [65]. This could be by posting overly negative or overly positive ratings maliciously.

- *Deceptive spam* happens when spammers create deceptive reviews, specifically to influence the perception of review readers. Such fake reviews can be created either manually [199] or automatically [188].

- *Group spam* happens when a group of spammers work in concert to create misleading reviews [191].

Both cognitive biases and spams exist in trust-based systems. Spams are malicious and clearly need to be detected and addressed. Cognitive biases, while not malicious, may still mislead in some cases. This chapter will not focus on identifying the *causes* (i.e., whether cognitive bias or spam, and which type). Rather, our focus is on reviewing various methods that have been proposed to *detect* and *lessen* the impact of biases.

6.3 Detection of Biases

Before bias can be appropriately addressed, we first need to be able to detect the occurrences of bias. This is a very challenging task because there are no clear-cut indicators of biases. In the case of cognitive bias, the perpetrators may not even be conscious of the various bias effects. In the case of spam, the perpetrators deliberately try to avoid detection.

There are two main approaches to detecting biases. The first is an unsupervised approach, which does not assume clear examples of biases. The second is a supervised approach, which assumes that some clear positive and negative examples are given so as to allow learning an appropriate classification or regression function.

6.3.1 Unsupervised Approaches

Statistical Deviation. One unsupervised approach is to identify "outliers". The presumption is that biased feedback (e.g., a rating or a review) represents only a relatively small subset of all feedbacks. Therefore, a feedback that appears very different from the others is worthy of further inspection.

This is an approach described in Lauw, Lim and Wang [133, 135], which outlines two deviation measures. The first deviation measure is called *deviation from the mean*, given by Equation 6.1. A rating e_{ij} by a reviewer r_i on an object o_j has a high deviation $d_{ij} \in [0, 1]$, if e_{ij} is very different from the average of all the ratings on o_j, i.e., $\{e_{kj}\}$. High deviation d_{ij} may be one indication of a possible bias on the part of reviewer r_i. This measure is simple and straightforward, but also has some disadvantages. For one thing, it unfairly favors those who rate close to the mean. For example, suppose r_1, r_2 and r_3 assign to o_j the following scores: $e_{1j} = 0.0$, $e_{2j} = 0.5$ and $e_{3j} = 1.0$, then we would have $d_{1j} = 0.5$, $d_{2j} = 0.0$, and $d_{3j} = 0.5$. It creates the impression that r_2 has not deviated at all, although clearly all three reviewers cannot agree with one another. This is due to the assumption that the average is close to being the "correct" score, which is not always a fair assumption.

$$d_{ij} = \left| e_{ij} - \underset{k}{Avg}\ e_{kj} \right| \tag{6.1}$$

The second deviation measure is called *deviation from co-reviewers*, given by Equation 6.2. Deviation d_{ij} is measured as the average distance between r_i's score and the score by each co-reviewer r_k, i.e., e_{kj}. m_j is the number of reviewers of o_j (including r_i). This measure captures the notion of disagreement from the majority. To have a low deviation, a reviewer needs to agree with most of the other reviewers. If inherently there is no agreement, no reviewer could have a low deviation score. For the same example of r_1, r_2 and r_3 assigning $e_{1j} = 0.0$, $e_{2j} = 0.5$ and $e_{3j} = 1.0$ to o_j, we have $d_{1j} = 0.75$,

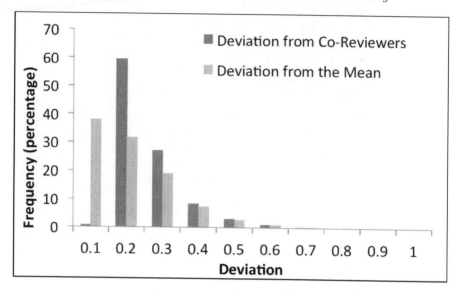

FIGURE 6.1: Distribution of deviation values for MovieLens.

$d_{2j} = 0.50$, and $d_{3j} = 0.75$. This is a more reasonable outcome than previously, as $d_{2j} = 0.50$ indicates that indeed r_2 is also showing some deviation from her co-reviewers.

$$d_{ij} = \begin{cases} \frac{1}{m_j-1} \sum_{k \neq i} |e_{kj} - e_{ij}| & \text{if } m_j > 1 \\ 0 & \text{otherwise} \end{cases} \qquad (6.2)$$

For an empirical comparison between the two deviation measures, Figure 6.1 plots a histogram of d_{ij} values obtained by applying the two measures on real-life data ("1 Million MovieLens Dataset"). The figure shows that for deviation from the mean, the resulting d_{ij} values are quite small; close to 40% are below 0.1; close to 70% are below 0.2. This may underestimate deviations of users close to the mean. In contrast, the d_{ij} values using deviation from co-reviewers on the same data have only a very small proportion ($< 1\%$) in the bin closest to zero (0.1 or less).

In any case, a deviation measure by itself is insufficient to conclude that a biased feedback has occurred. An instance of high deviation could simply be a statistical aberration. For a greater confidence, we need repeated instances of deviation. It is therefore an easier problem to associate bias with a user or a reviewer, rather than with individual ratings. This is because a reviewer would have a history of deviations from past ratings. One way to infer how biased a reviewer is is by aggregating her deviations. As shown in Equation 6.3, the bias $b_i \in [0, 1]$ of a reviewer r_i can be expressed as the average of deviations d_{ij} on various objects o_j's:

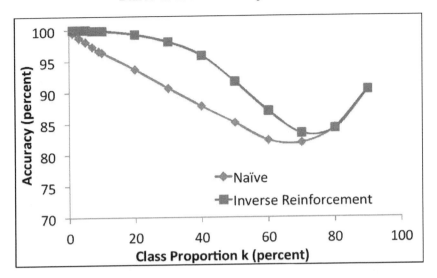

FIGURE 6.2: Bias: Varying proportion of bias class.

$$b_i = \underset{j}{Avg} \ d_{ij} \qquad (6.3)$$

The above equation is simple, but has not taken into account other factors that may affect the computation of bias. For instance, Lauw et al. [133, 135] link the bias of reviewers to the controversy of objects, such that a reviewer who repeatedly deviates on controversial objects (that have no agreement among reviewers) may not be biased. Equation 6.4 shows how in computing the bias b_i, r_i's deviation d_{ij} on an object o_j is "discounted" by the controversy value $c_j \in [0, 1]$ of the object: The more controversial the object is (i.e., higher c_j), the less d_{ij} will contribute to the bias b_i.

$$b_i = \underset{j}{Avg} \ d_{ij} \cdot (1 - c_j) \qquad (6.4)$$

In turn, to quantify the controversy c_j, there is a need to pay attention to the deviations by various reviewers on o_j. Using a similar principle, c_j is computed based on Equation 6.5, whereby the bias of a reviewer is also taken into account when determining the controversy. This creates an *inverse reinforcement* relationship between b_i and o_j of all reviewers and items in the system, which requires these quantities to be resolved simultaneously through an iterative computation.

$$c_j = \underset{i}{Avg} \ d_{ij} \cdot (1 - b_i) \qquad (6.5)$$

While Lauw et al. [133, 135] experimented with both real-life and synthetic datasets, this chapter covers a couple of their key results on the synthetic

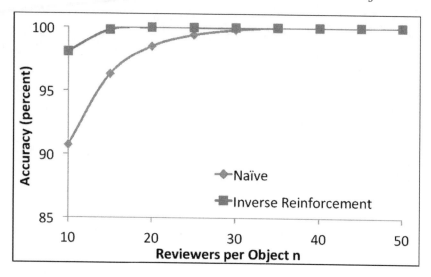

FIGURE 6.3: Bias: Varying reviewers per object.

dataset. The synthetic dataset is created by simulating the behavior of two classes of reviewers. The first class (biased reviewers) would introduce random ratings. The second class (unbiased reviewers) would give "true" ratings. Given the ratings alone, the bias b_i of each reviewer can be computed. Starting with k percent of the reviewers belonging to the first class, the top k percent of reviewers based on their highest bias values are classified as biased. It is then feasible to compare the accuracies by the *Naive* method (Equation 6.3) and the *Inverse Reinforcement* method (Equations 6.4 and 6.5).

Figure 6.2 plots the accuracies varying with the proportion of the biased reviewers, where every object is assigned 10 reviewers randomly. Overall, the accuracies would go down as k increases. A higher proportion of biased reviewers and controversial objects means more propagation effects due to bias and controversy in the network. It becomes easier to confuse a biased reviewer, with a reviewer who simply has evaluated controversial objects. The anomalous "uptick" for $k \geq 80\%$ is simply due to the majority effect. For instance, at $k = 90\%$, the accuracy is also 90%, implying that a simple majority classifier could achieve the same accuracy. Importantly, the *Inverse Reinforcement* method has higher accuracies than the *Naive* method across different k's.

Figure 6.3 plots the accuracies when every object is assigned an increasing number of reviewers n, for $k = 30\%$. It shows that the more reviewers are involved, the easier it becomes to identify the biased users, as those types of behaviors stand out more clearly. Importantly, when there are few reviewers involved (e.g., $n < 20$), the *Inverse Reinforcement* method does a significantly better job than the *Naive* at identifying biased users.

While the above discussion links bias to controversy, there are also other works (e.g., Mishra and Rastogi [182]) linking bias to a more accurate estimation of the quality of objects.

Specific Bias Patterns. The study of deviation simply accounts for some "unexpected" or "outlier" behavior, without any specific hypothesis of the pattern in the biased behavior. In some cases, we may be looking for a specific pattern of behavior, which indicates one of the biases mentioned in Section 6.2. Here is an outline of two examples of bias pattern.

1. One such behavior pattern is *leniency* [134, 136], which is the tendency to systematically overrate (or underrate). Suppose that the true quality of an object q_j is known. The leniency l_i of a reviewer is then measured as a function of how much the given score e_{ij} is inflated or deflated, i.e., $e_{ij} - q_j$, as shown in Equation 6.6. The more positive the value, the more lenient the reviewer is. The more negative, the less lenient she is. In some cases, it may be desired to express leniency in relative terms, in which case it can be normalized by either the rating score e_{ij} or the quality q_j [134, 136]. The more regularly a reviewer r_i inflates her rating scores, the more evidence there is that r_i is lenient. This explains why Equation 6.6 takes the average across all objects rated by the same reviewer.

$$l_i = \underset{j}{Avg} \left(e_{ij} - q_j \right) \tag{6.6}$$

2. Another approach to discovering bias is by looking for *suspicious patterns of association* [250]. For example, in terms of academic peer review, it is well recognized that an academic advisor and an advisee should not be reviewing each other's submissions. In terms of product reviews, e.g., book reviews, it may raise some flags if two book authors tend to review each other's books. While there may be a limited number of types of such "biased" associations, identifying all such instances is difficult due to the scale of the data. Staddon and Chow [250] presented an approach to identify such associations efficiently through association rule mining [12].

Because these approaches are unsupervised, they may not conclusively detect biases. These approaches are, however, useful to narrow the search space, by identifying "potential" instances of bias, which can be further examined by domain experts to establish whether they are genuine bias or simply statistical anomalies.

6.3.2 Supervised Approaches

A second direction in detecting biases makes use of supervised approaches. This approach is useful when it is possible to identify clear evidence of biased behavior, which can then be used as training labels. A supervised approach identifies a number of class labels, which may correspond to certain types of biased behavior.

Most work in this area tackles spam, rather than cognitive biases. This is because spam is somewhat easier to identify to be used as labels. Moreover, supervised approaches, to a large extent, rely on the availability of good features. Reviews (being textual in nature) tend to generate a richer amount of features than ratings (being numerical in nature). For this reason, our discussion of supervised detection of bias will focus on the works on spam reviews.

The general approach is thus to generate a set of training labels (both positive and negative), which are then fed into a classification or regression algorithm, using a number of features. In the following, let us look at several ways in which labels and features can be generated.

Labels. To train a classifier for two classes: *spam* vs. *non-spam*, there is a need to generate some reviews of each class label. However, genuine spammers would take pains to hide their traces, which makes it difficult to identify the genuine cases of spam reviews. Several approaches have been discussed in the literature.

- *Manual.* One approach to generate spam reviews to be used as labels is to ask human subjects to write fake reviews. This approach was employed by Ott, Choi, Cardie and Hancock [199], which used Amazon Mechanical Turk to generate 400 fake reviews for 20 hotels in Chicago. These fake reviews were then used as positive examples (spam). In turn, the negative examples (non-spam) were obtained by sampling reviews from TripAdvisor.com.

- *Synthetic.* Another approach to generate spam reviews is to design computer algorithms to do so. Morales, Sun and Yan [188] generated a fake review by starting from an original review that was used as a base "template". The algorithm then replaced each sentence in this base review with another sentence of similar content from the corpus of reviews. The result was a fake review that made use of real sentences from other reviews. These synthesized reviews were then used as positive examples, while negative examples were again obtained by sampling base reviews.

- *Reverse-engineering.* Yet another approach is to reverse-engineer another spam identification system. For instance, Mukherjee et al. [192] used Yelp's filtered reviews as positive examples. These are reviews deemed suspicious by the restaurant review site Yelp.com. Correspondingly, the negative examples could be drawn from other nonfiltered reviews in Yelp.

Note that none of the above label generation approaches is really ideal, because the positive examples may not be actual genuine fake reviews generated by real spammers. Nevertheless, each approach allows a study of the useful set of features.

Features. To train the classifier, there is a need to identify a set of features from reviews. Various features have been explored in the literature, including the following.

- *Lexical.* Because reviews are textual, one category of features that is readily available are the words themselves. The assumption is that spam reviews may contain certain words with different frequencies from original reviews. In particular, Ott et al. [199] considers N-grams as features, for example unigrams, bigrams and trigrams.

- *Syntactic.* Another category of features is syntactic features. The assumption is that spam reviews have sentence structures that are different from original reviews. Feng, Banerjee and Choi [64] considered both shallow syntax, such as part-of-speech tags, as well as deep syntax in the form of production rules based on Probabilistic Context Free Grammar.

- *Temporal.* Reviews also frequently have timestamps. Some works hypothesize that the timing of some reviews may indicate deception. For instance, Fei et al. [63] studied bursts in reviews, which may indicate either sudden popularity or spam attacks. In another study, Lim et al. [145] included "early deviation", or how soon a review deviates from the others, as a feature.

- *Behavioral.* Other than the reviews, the nature of the *review writers* may also be informative. Mukherjee et al. [190] listed a number of user behaviors as features. This includes the number of reviews written by the user, whether they tend to be similar to other reviews, whether they tend to have extreme ratings or deviating ratings, etc.

- *Coherence.* A clue to the originality of a review is how consistent the different parts of the review are. A fake review, especially one that is created by copying or synthesizing from multiple other reviews, would tend to have parts that are not coherent [188]. Among other things, coherence may be indicated by the smoothness of transition between consecutive sentences.

The above categories of features have been useful in several instances of spam reviews classification. The focus of these works is mostly on the set of labels and features, but not on the specific classification algorithm. Most works make use of well-known classification algorithms such as Naive Bayes [173] and Support Vector Machine [43].

6.4   Lessening the Impact of Biases

Because detection of bias is such a hard task, the impact of bias probably cannot be removed completely. However, it can be lessened through various strategies. In this discussion, let us assume that the task at hand is to estimate the "true" quality q_j of an object o_j based on the rating scores e_{ij} assigned by various reviewers r_i's.

6.4.1   Avoidance

The first strategy is to avoid having "biased" rating scores in the first place. This can be done for instance by ensuring that people with vested interests or known possible biases refrain from assigning a rating score. There are examples of this strategy in various settings. In academic peer review, authors making a submission are frequently requested to disclose the list of potential reviewers with conflicts of interests, e.g., advisor, advisee, collaborator. That way, such conflicted reviewers would not be asked to give their opinions, avoiding potentially biased opinions.

The advantage of avoidance strategy is that it presents a relatively clean set of rating scores, provided those with potential vested interests can be identified accurately. However, its very conservative approach also has a disadvantage in reducing the pool of potential reviewers. Another disadvantage is that it may not be applicable to all scenarios. For instance, in online settings, literally anyone can submit a rating or a review. Since people hide behind pseudonyms or user names, it is very difficult to enforce who may or may not assign a rating score.

6.4.2   Aggregation

The second strategy is to consider all the rating scores, but not to trust any one individual score completely. Instead all the reviewer scores for the same object are aggregated into a single number that reflects the quality of the object. Provided there are sufficiently many reviewers and the reviewers assign rating scores independently, the aggregation probably would even out the various biases and produce a relatively unbiased estimate of the quality. An instance of aggregation function based on average is shown in

$$q_j = \underset{i}{Avg}\ e_{ij} \tag{6.7}$$

In most cases, there are not that many reviewers involved in rating a particular object. It may be necessary to assign different weights to the contributions of various reviewers [48]. Let us denote with $w_i \in \mathbb{R}_0^+$ the weight of reviewer r_i's contribution, with a higher w_i indicating a greater "trust" in r_i.

The overall quality q_j can then be expressed as a weighted average of rating scores e_{ij}, weighted by w_i:

$$q_j = \frac{\sum_i w_i \times e_{ij}}{\sum_i w_i} \tag{6.8}$$

The complexity is then in determining w_i. In a rating system, each reviewer r_i usually has established a history of previous ratings. From this historical data, it is feasible to estimate the appropriate w_i.

Here are two examples of how w_i may be derived.

- In Riggs and Wilensky [223], w_i is based on consensus, i.e., how closely r_i generally assigns a rating score that is close to the estimated quality of objects, as shown in Equation 6.9. The underlying principle is that a more trusted reviewer is one who has mostly been "right" previously (with respect to the consensus, as the true quality is not known).

$$w_i = 1 - \underset{j}{Avg} \, |e_{ij} - q_j| \tag{6.9}$$

- In Mishra and Rastogi [182], w_i is based on the detected bias b_i (see Section 6.3) of the reviewer r_i. As shown in Equation 6.10, a more trusted reviewer is one who has small detected bias.

$$w_i = 1 - b_i \tag{6.10}$$

There are also other ways to estimate the weight w_i. Wei et al. [283] considered both consistent inaccuracies and inconsistent inaccuracies. Chen and Singh [37] considered the opinions of other reviewers in deriving the weight.

An advantage of this strategy is that it employs an intelligent strategy to decide whom to trust or distrust. However, the weighted average in Equation 6.8 means that the approach will work only if there are a number of "trusted" reviewers with high weight w_i rating an object. In that case, the aggregated quality will be close to the ratings of these reviewers.

On the other hand, many times there are only a few reviewers for any one object. This increases the chance that a particular object may receive rating scores from only reviewers with low weights w_i. Equation 6.8 still cannot "correct" for the (supposedly biased) ratings from reviewers with low weights in computing q_j. This motivates the need for a compensation approach that can partially "correct" a biased rating.

6.4.3 Compensation

In contrast to the aggregation strategy, which takes each rating score as it is, the compensation strategy seeks to compensate for the "bias" directly. This is feasible only if there is a systematic direction in the bias. To illustrate this compensation strategy, let us use the same notion of leniency as defined in Equation 6.6 in Section 6.3.

For instance, suppose a reviewer r_i has a leniency of $l_i = 0.2$. What this suggests is that this reviewer may systematically inflate her rating score by about 0.2. To bring her rating score e_{ij} down to *compensate* for this inflation, one approach is to reduce e_{ij} by a proportional amount to l_i. The quality q_j of an object can thus be computed according to Equation 6.11, where $\alpha \in [0, 1]$ denotes the extent of compensation. It follows that if $l_i < 0$, the compensation will go in the opposite direction, i.e., to bring up the rating score because in this case the reviewer tends to deflate her scores.

$$q_j = \underset{i}{Avg} \left(e_{ij} - \alpha \cdot l_i \right) \qquad (6.11)$$

Note that Equation 6.6 and Equation 6.11 are mutually dependent because they are expressed in terms of each other. Lauw et al. [134, 136] described algorithms based on inverse reinforcement to resolve this mutual dependency.

To see how the leniency-based compensation of quality (Equations 6.11 and 6.6) performs in comparison to the aggregation strategy, let us look at a subset of the experimental results in Lauw et al. [134, 136]. The two aggregation methods are (1) the *Naive* method (Equation 6.7), and (2) the weighted aggregation method [223] (Equations 6.8 and 6.9), referred to as the *Riggs* method after its first author.

Since the "true" quality is usually unknown, to investigate the differences among methods, let us use a synthetic dataset generated as follows. Each object is randomly assigned a "true" quality. Among the set of reviewers, 30% are classified as lenient, i.e., they will assign a random rating higher than the "true" rating. Another 30% are strict, i.e., they will assign a random rating lower than the "true" rating. The remainder 40% are neutral, i.e., they will assign the "true" rating. Every object is then assigned n reviewers drawn randomly from the above population of reviewers. The next step is to use the three methods (*Naive*, *Riggs* and *leniency-based*) to compute the quality of each object. The metric is the Kendall rank correlation coefficient [5], which is used to compare two ranking lists. Each method's ranking of objects' qualities is compared to the "ground-truth" quality ranking. The higher the Kendall rank correlation score (up to 100%), the closer is the quality ranking to the "ground truth".

Figure 6.4 shows how the Kendall rank correlation varies for different number of reviewers assigned to every object n, for $\alpha = 0.7$. As n increases, all three methods have better correlation to the ground truth. This is because greater connectivity between reviewers and objects makes the problem of quality determination easier, as each object tends to get similar proportion of lenient and strict reviewers, which even out their effects. The results show that *Riggs* outperforms *Naive*, which suggest that weighted aggregation works when there are a good weights for the reputation of the reviewers. Moreover,

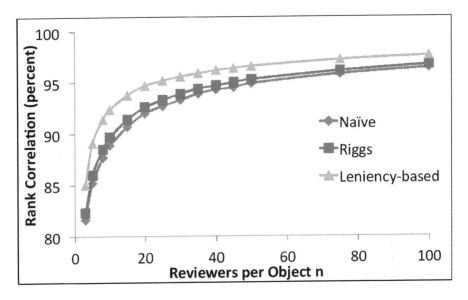

FIGURE 6.4: Quality: Varying reviewers per object.

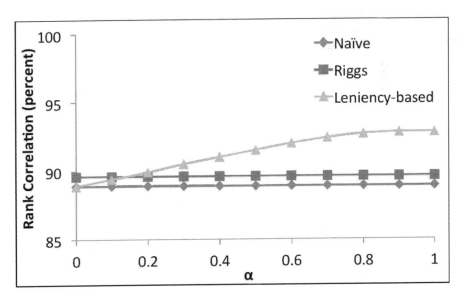

FIGURE 6.5: Quality: Varying compensation factor α.

the *leniency-based* method outperforms both aggregation methods across all values of n.

Figure 6.5 shows the effect of the compensation factor α in Equation 6.11. The Kendall rank correlation tends to increase with α. We see that the greater the α, the further away (the better) is the leniency-based method as compared to the aggregation-based *Naive* and *Riggs* methods, due to the greater degree of compensation. Roos et al. [225, 226] proposed to estimate parameters such as α from data.

6.4.4 Elimination

Sometimes the appropriate course of action is to eliminate biased feedback from consideration. This is especially the case when the biased feedback is due to malicious or deceptive intent, such as spam reviews. This strategy is actually practiced in industry. For instance, Yelp.com, a restaurant review site, has actively been identifying suspicious reviews, removing them from the main content, and isolating them to a special section called "filtered reviews" [192]. This way, such biased content will not affect the experience of the users.

The caveat to this approach is the accuracy (or lack thereof) of the bias detection algorithm. The higher the confidence in bias detection, the simpler the biased feedback elimination would be. Otherwise, there is a risk of erroneously removing good feedbacks that are wrongly classified as biased.

6.5 Summary

Bias affects many trust-based systems, such as ratings and reviews. Some of these instances are due to various types of cognitive biases, which are natural to human subjects. Others are due to malice by individuals out to deceive. In specific settings, some occurrences of bias can be detected, either through unsupervised approaches based on deviation and outlier analysis, or through supervised approaches based on classification and regression. Once specific occurrences of bias are detected or identified, there are several ways to address them. A simple strategy is to avoid them altogether. Where not possible, coping strategies include relying on aggregation or compensation to "correct" the biased feedback. When there is high confidence in the bias detection, the biased feedback may be eliminated altogether. Bias detection and bias correction are active areas of study with further opportunities for advancement.

Glossary

Cognitive Bias: Deviation in judgments that affects human thinking processes.

Spam: Disruptive or deceptive content.

Acknowledgments

Most of the quantitative results presented in this chapter were produced as part of a research collaboration. Where appropriate, specific attributions are provided by citing the original articles in journals and conference proceedings where these results first appeared.

In addition, the author would like to acknowledge the support of A*STAR and Nanyang Technological University in the course of research investigation leading to some of the above-mentioned results, as well as the support of Singapore Management University in the course of writing this book chapter.

Bibliography

[1] Do consumers still believe what is said in online product reviews? A persuasion knowledge approach. *Journal of Retailing and Consumer Services*, 20(4):373 – 381, 2013.

[2] Mohammad-Ali Abbasi, Sun-Ki Chai, Huan Liu, and Kiran Sagoo. Real-world behavior analysis through a social media lens. In *Proceedings of the 5th International Conference on Social Computing, Behavioral-Cultural Modeling and Prediction*, SBP'12, pages 18–26. College Park, MD, 2012.

[3] Mohammad-Ali Abbasi and Huan Liu. Measuring user credibility in social media. In *Proceedings of the 6th International Conference on Social Computing, Behavioral-Cultural Modeling and Prediction*, SBP'13, pages 441–448, Washington, DC, 2013.

[4] Zeinab Abbassi, Christina Aperjis, and Bernardo A Huberman. Friends versus the crowd: Tradeoffs and dynamics. *HP Report*, 2013.

[5] Hervé Abdi. The kendall rank correlation coefficient. In Neil J. Salkind, editor, *Encyclopedia of Measurement and Statistics*, pages 508–510. Sage, Thousand Oaks, CA, 2007.

[6] Alfarez Abdul-Rahman and Stephen Hailes. A distributed trust model. In *Proceedings of the 1997 Workshop on New Security Paradigms (NSPW)*, pages 48–60, Langdale, Cumbria, UK, 1997.

[7] Alfarez Abdul-Rahman and Stephen Hailes. Supporting trust in virtual communities. In *Proceedings of the 33rd Hawaii International Conference on System Sciences*, HICSS '00, page 6007, Maui, Hawaii, 2000.

[8] Karl Aberer and Zoran Despotovic. Managing trust in a peer-2-peer information system. In *Proceedings of the Tenth International Conference on Information and Knowledge Management*, CIKM '01, pages 310–317, Atlanta, Georgia, 2001.

[9] B. Thomas Adler, Krishnendu Chatterjee, Luca de Alfaro, Marco Faella, Ian Pye, and Vishwanath Raman. Assigning trust to wikipedia content. In *Proceedings of the 4th International Symposium on Wikis*, WikiSym '08, pages 26:1–26:12, Porto, Portugal, 2008.

[10] B. Thomas Adler and Luca de Alfaro. A content-driven reputation system for the wikipedia. In *Proceedings of the 16th International Conference on World Wide Web*, WWW '07, pages 261–270, Banff, Alberta, Canada, 2007.

[11] Gediminas Adomavicius and Alexander Tuzhilin. Toward the next generation of recommender systems: A survey of the state-of-the-art and possible extensions. *Knowledge and Data Engineering, IEEE Transactions on*, 17(6):734–749, 2005.

[12] Rakesh Agrawal and Ramakrishnan Srikant. Fast algorithms for mining association rules in large databases. In *Proceedings of the 20th International Conference on Very Large Data Bases*, VLDB '94, pages 487–499, Santiago de Chile, Chile, 1994.

[13] Donovan Artz and Yolanda Gil. A survey of trust in computer science and the semantic web. *Web Semant.*, 5(2):58–71, 2007.

[14] B P Bailey, L J Gurak, and J A Konstan. An examination of trust production in computer-mediated exchange. In *Proceedings of the 7th Conference on Human Factors and the Web*, 2001.

[15] Marko Balabanović and Yoav Shoham. Fab: Content-based, collaborative recommendation. *Communications of the ACM*, 40(3):66–72, 1997.

[16] Elisa Bertino, Chenyun Dai, Hyo-Sang Lim, and Dan Lin. High-assurance integrity techniques for databases. In *Proceedings of the 25th British National Conference on Databases: Sharing Data, Information and Knowledge*, BNCOD '08, pages 244–256, Cardiff, Wales, UK, 2008.

[17] J Patrick Biddix, Chung Joo Chung, and Han Woo Park. Convenience or credibility? A study of college student online research behaviors. *The Internet and Higher Education*, 14(3):175–182, 2011.

[18] Hubert M Blalock. *Social statistics revised*. McGraw-Hill, 2 edition, 1979.

[19] Gary Bolton, Ben Greiner, and Axel Ockenfels. Engineering trust: Reciprocity in the production of reputation information. *Management Science*, 59(2):265–285, 2013.

[20] John S. Breese, David Heckerman, and Carl Kadie. Empirical analysis of predictive algorithms for collaborative filtering. In *Proceedings of the Fourteenth Conference on Uncertainty in Artificial Intelligence*, UAI'98, pages 43–52, Madison, Wisconsin, 1998.

[21] Percy Williams Bridgman. *Dimensional Analysis*. New Haven : Yale University Press, 1922.

[22] S Brin and L Page. The anatomy of a large-scale hypertextual Web search engine. *Computer Networks and ISDN Systems*, 30(1–7):107–117, 1998.

[23] Sonja Buchegger and Jean-Yves Le Boudec. The effect of rumor spreading in reputation systems for mobile ad-hoc networks. In *Proceedings of WiOpt 03: Modeling and Optimization in Mobile, Ad Hoc and Wireless Networks*, Sophia-Antipolis, France, 2003.

[24] Erik P Bucy. Media credibility reconsidered: Synergy effects between on-air and online news. *Journalism & Mass Communication Quarterly*, 80(2):247–264, 2003.

[25] David B Buller and Judee K Burgoon. Interpersonal deception theory. *Communication Theory*, 6(3):203–242, 1996.

[26] Robin Burke. Hybrid recommender systems: Survey and experiments. *User Modeling and User-Adapted Interaction*, 12(4):331–370, 2002.

[27] Robin Burke. The adaptive web. chapter Hybrid Web Recommender Systems, pages 377–408. Springer-Verlag, Berlin, Heidelberg, 2007.

[28] Chris Burnett, Timothy J. Norman, and Katia Sycara. Bootstrapping trust evaluations through stereotypes. In *Proceedings of the 9th International Conference on Autonomous Agents and Multiagent Systems*, pages 241–248, Toronto, Canada, 2010.

[29] Chris Burnett, Timothy J. Norman, and Katia Sycara. Sources of stereotypical trust in multi-agent systems. In *Proceedings of the Fourteenth International Workshop on Trust in Agent Societies*, pages 25–39, Taipei, Taiwan, 2011.

[30] Vincent Buskens. The social structure of trust. *Social Networks*, 20(3):265–289, 1998.

[31] Andrea Caputo. Relevant information, personality traits and anchoring effect. *International Journal of Management and Decision Making*, 13(1):62–76, 2014.

[32] Andrew Carlson, Justin Betteridge, Bryan Kisiel, Burr Settles, Estevam Hruschka, and Tom Mitchell. Toward an architecture for never-ending language learning. In *Proceedings of the Conference on Artificial Intelligence (AAAI)*, pages 1306–1313, Atlanta, Georgia, 2010.

[33] Christiano Castelfranchi and Rino Falcone. *Trust theory: A Socio-Cognitive and Computational Model*, volume 18. John Wiley & Sons, Hoboken, NJ, 2010.

[34] Carlos Castillo, Marcelo Mendoza, and Barbara Poblete. Information credibility on twitter. In *Proceedings of the 20th International Conference on World Wide Web*, pages 675–684, Hyderabad, India, 2011.

[35] James Caverlee, Ling Liu, and Steve Webb. Socialtrust: Tamper-resilient trust establishment in online communities. In *Proceedings of the 8th ACM/IEEE-CS Joint Conference on Digital Libraries*, pages 104–114, Pittsburgh, PA, 2008.

[36] Ming-Wei Chang, Lev Ratinov, and Dan Roth. Structured learning with constrained conditional models. *Mach. Learn.*, 88(3):399–431, September 2012.

[37] Mao Chen and Jaswinder Pal Singh. Computing and using reputations for Internet ratings. In *Proceedings of the 3rd ACM Conference on Electronic Commerce*, EC '01, pages 154–162, Tampa, Florida, 2001.

[38] Paul-Alexandru Chirita, Wolfgang Nejdl, and Cristian Zamfir. Preventing shilling attacks in online recommender systems. In *Proceedings of the 7th Annual ACM International Workshop on Web Information and Data Management*, WIDM '05, pages 67–74, Bremen, Germany, 2005.

[39] Jinsook Cho. The mechanism of trust and distrust formation and their relational outcomes. *Journal of Retailing*, 82(1):25–35, 2006.

[40] Jacob Cohen, Patricia Cohen, Stephen G West, and Leona S Aiken. *Applied Multiple Regression/Correlation Analysis for the Behavioral Sciences*. Routledge, London, UK, 2013.

[41] William Conner, Arun Iyengar, Thomas Mikalsen, Isabelle Rouvellou, and Klara Nahrstedt. A trust management framework for service-oriented environments. In *Proceedings of the 18th International Conference on World Wide Web*, pages 891–900, Madrid, Spain, 2009.

[42] Jonathan R Corney, Carmen Torres-Sánchez, A Prasanna Jagadeesan, and William C Regli. Outsourcing labour to the cloud. *International Journal of Innovation and Sustainable Development*, 4(4):294–313, 2009.

[43] Corinna Cortes and Vladimir Vapnik. Support vector machine. *Machine Learning*, 20(3):273–297, 1995.

[44] Chenyun Dai, Dan Lin, Elisa Bertino, and Murat Kantarcioglu. An approach to evaluate data trustworthiness based on data provenance. In *Proceedings of the 5th VLDB Workshop on Secure Data Management*, SDM '08, pages 82–98, Auckland, New Zealand, 2008.

[45] Chenyun Dai, Dan Lin, Elisa Bertino, and Murat Kantarcioglu. Trust evaluation of data provenance. Technical report, Purdue University, West Lafayette, IN, 2008.

[46] Ernesto Damiani, De Capitani di Vimercati, Stefano Paraboschi, Pierangela Samarati, and Fabio Violante. A reputation-based approach for choosing reliable resources in peer-to-peer networks. In *Proceedings of the 9th ACM Conference on Computer and Communications Security*, CCS '02, pages 207–216, Washington, DC, 2002.

[47] Alexander Philip Dawid and Allan M Skene. Maximum likelihood estimation of observer error-rates using the em algorithm. *Applied Statistics*, 28:20–28, 1979.

[48] Morris H DeGroot. Reaching a consensus. *Journal of the American Statistical Association*, 69(345):118–121, 1974.

[49] Dominik Deja. Crowdsourcing in scientific experiments. Master's thesis, Warsaw School of Economics, 2013.

[50] Chrysanthos Dellarocas. Immunizing online reputation reporting systems against unfair ratings and discriminatory behavior. In *Proceedings of the 2Nd ACM Conference on Electronic Commerce*, EC '00, pages 150–157, Minneapolis, Minnesota, 2000.

[51] Chrysanthos Dellarocas and Charles A Wood. The sound of silence in online feedback: Estimating trading risks in the presence of reporting bias. *Management Science*, 54(3):460–476, 2008.

[52] Arthur P Dempster, Nan M Laird, and Donald B Rubin. Maximum likelihood from incomplete data via the em algorithm. *Journal of the Royal Statistical Society. Series B (Methodological)*, 39:1–38, 1977.

[53] Daniel C. Dennett. Postmodernism and truth. In *Proceedings of the Twentieth World Congress of Philosophy*, volume 8, pages 93–103, 2000.

[54] Zoran Despotovic and Karl Aberer. P2P reputation management: Probabilistic estimation vs. social networks. *Computer Networks*, 50(4):485–500, March 2006.

[55] Chris H. Q. Ding, Tao Li, and Michael I. Jordan. Nonnegative matrix factorization for combinatorial optimization: Spectral clustering, graph matching, and clique finding. In *Proceedings of the 13th International Conference on Data Mining*, pages 183–192, Pisa, Italy, 2008.

[56] Yi Ding and Xue Li. Time weight collaborative filtering. In *Proceedings of the 14th ACM International Conference on Information and Knowledge Management*, pages 485–492, Bremen, Germany, 2005.

[57] Xin Luna Dong, Laure Berti-Equille, and Divesh Srivastava. Integrating conflicting data: The role of source dependence. *Very Large DataBases Endowment*, 2(1):550–561, 2009.

[58] Xin Luna Dong, Laure Berti-Equille, and Divesh Srivastava. Truth discovery and copying detection in a dynamic world. *Very Large DataBases Endowment*, 2(1):562–573, 2009.

[59] Julie S. Downs, Mandy B. Holbrook, Steve Sheng, and Lorrie Faith Cranor. Are your participants gaming the system?: Screening mechanical turk workers. In *Proceedings of the SIGCHI Conference on Human Factors in Computing Systems*, CHI '10, pages 2399–2402, Atlanta, Georgia, 2010.

[60] Susan T. Dumais and Jakob Nielsen. Automating the assignment of submitted manuscripts to reviewers. In *Proceedings of the 15th Annual International ACM SIGIR Conference on Research and Development in Information Retrieval*, SIGIR '92, pages 233–244, Copenhagen, Denmark, 1992.

[61] Ehab ElSalamouny, Vladimiro Sassone, and Mogens Nielsen. Hmm-based trust model. In Pierpaolo Degano and Joshua D. Guttman, editors, *Formal Aspects in Security and Trust*, volume 5983 of *Lecture Notes in Computer Science*, pages 21–35. Springer, Berlin Heidelberg, 2010.

[62] Yi Fang and Luo Si. Matrix co-factorization for recommendation with rich side information and implicit feedback. In *Proceedings of the 2nd International Workshop on Information Heterogeneity and Fusion in Recommender Systems*, HetRec '11, pages 65–69, Chicago, Illinois, 2011.

[63] Geli Fei, Arjun Mukherjee, Bing Liu, Meichun Hsu, Malu Castellanos, and Riddhiman Ghosh. Exploiting burstiness in reviews for review spammer detection. In *Proceedings of the 7th International AAAI Conference on Weblogs and Social Media*, Ann Arbor, Michigan, 2013.

[64] Song Feng, Ritwik Banerjee, and Yejin Choi. Syntactic stylometry for deception detection. In *Proceedings of the 50th Annual Meeting of the Association for Computational Linguistics*, ACL '12, pages 171–175, Jeju Island, Korea, 2012.

[65] Song Feng, Longfei Xing, Anupam Gogar, and Yejin Choi. Distributional footprints of deceptive product reviews. In *Proceedings of the 6th International AAAI Conference on Weblogs and Social Media*, Dublin, Irland, 2012.

[66] Andrew J Flanagin and Miriam J Metzger. The credibility of volunteered geographic information. *GeoJournal*, 72(3-4):137–148, 2008.

[67] B. J. Fogg. Prominence-interpretation theory: Explaining how people assess credibility online. In *CHI '03 Extended Abstracts on Human Factors in Computing Systems*, CHI EA '03, pages 722–723, Ft. Lauderdale, Florida, 2003.

[68] B J Fogg, J Marshall, T Kameda, J Solomon, A Rangnekar, J Boyd, and B Brown. Web credibility research: A method for online experiments and early study results. In *Proceedings of ACM CHI 2001 Conference on Human Factors in Computing Systems*, pages 295–296, Seattle, Washington, 2001.

[69] B J Fogg, P Swani, M Treinen, J Marshall, O Laraki, A Osipovich, C Varma, N Fang, J Paul, A Rangnekar, and Others. What makes Web sites credible?: A report on a large quantitative study. In *Proceedings of ACM CHI 2001 Conference on Human Factors in Computing Systems*, pages 61–68, Seattle, Washington, 2001.

[70] B. J. Fogg and Hsiang Tseng. The elements of computer credibility. In *Proceedings of the SIGCHI Conference on Human Factors in Computing Systems*, CHI '99, pages 80–87, Pittsburgh, Pennsylvania, 1999.

[71] Pew Research Center for the People & the Press. Online papers modestly boost newspaper readership: Maturing Internet news audience broader than deep. *http://www.peoplepress.org/reports*, July 30 2006.

[72] Batya Friedman and Peter H. Kahn, Jr. The human-computer interaction handbook. chapter Human Values, Ethics, and Design, pages 1177–1201. L. Erlbaum Associates Inc., Hillsdale, NJ, 2003.

[73] John W Fritch. Heuristics, tools, and systems for evaluating Internet information: Helping users assess a tangled web. *Online Information Review*, 27(5):321–327, 2003.

[74] Alban Galland, Serge Abiteboul, Amélie Marian, and Pierre Senellart. Corroborating information from disagreeing views. In *Proceedings of the Third ACM International Conference on Web Search and Data Mining*, WSDM '10, pages 131–140, New York, NY, 2010.

[75] Francis Galton. Vox populi (the wisdom of crowds). *Nature*, 75(1949):450–451, 1907.

[76] Diego Gambetta. Can we trust trust? In Diego Gambetta, editor, *Trust: Making and Breaking Cooperative Relations*, pages 213–237. Basil Blackwell, Oxford, England, UK, 1988.

[77] Florent Garcin, Boi Faltings, Radu Jurca, and Nadine Joswig. Rating aggregation in collaborative filtering systems. In *Proceedings of the Third ACM Conference on Recommender Systems*, RecSys '09, pages 349–352, New York, NY, USA, 2009.

[78] Laura Garton, Caroline Haythornthwaite, and Barry Wellman. Studying online social networks. *Journal of Computer-Mediated Communication*, 3(1):0, 1997.

[79] Y Gil and D Artz. Towards content trust of Web resources. *Web Semantics: Science, Services and Agents on the World Wide Web*, 5(4):227–239, December 2007.

[80] Jim Giles. Internet encyclopaedias go head to head. *Nature*, 438(7070):900–901, 2005.

[81] Jennifer Golbeck and James Hendler. Filmtrust: Movie recommendations using trust in web-based social networks. In *Proceedings of the IEEE Consumer Communications and Networking Conference*, volume 96, pages 282–286, Las Vegas, Nevada, 2006.

[82] Jennifer Golbeck and James Hendler. Inferring binary trust relationships in web-based social networks. *ACM Transactions on Internet Technology*, 6(4):497–529, 2006.

[83] Jennifer Ann Golbeck. *Computing and Applying Trust in Web-based Social Networks*. PhD thesis, College Park, MD, USA, 2005.

[84] R. T. Golembiewski and M. McConkie. The centrality of interpersonal trust in group processes. *Theories of Group Processes*, pages 131–185, 1975.

[85] Marco Gori and Ian Witten. The bubble of web visibility. *Communications of the ACM*, 48(3):115–117, 2005.

[86] Nathan Griffiths. Task delegation using experience-based multi-dimensional trust. In *Proceedings of the Fourth International Joint Conference on Autonomous Agents and Multiagent Systems*, AAMAS '05, pages 489–496, Utrecht, Netherlands, 2005.

[87] R. Guha, Ravi Kumar, Prabhakar Raghavan, and Andrew Tomkins. Propagation of trust and distrust. In *Proceedings of the 13th International Conference on World Wide Web*, WWW '04, pages 403–412, New York, NY, 2004.

[88] Albert C. Gunther. *Extremity of Attitude and Trust in Media News Coverage of Issues*. PhD thesis, Stanford University, 1987.

[89] Albert C Gunther. Biased press or biased public? Attitudes toward media coverage of social groups. *Public Opinion Quarterly*, 56(2):147–167, 1992.

[90] Guibing Guo, Jie Zhang, Daniel Thalmann, Anirban Basu, and Neil Yorke-Smith. From ratings to trust: An empirical study of implicit trust in recommender systems. In *Proceedings of the 29th Annual ACM Symposium on Applied Computing*, Gyeongju, Korea, 2014.

[91] Aditi Gupta and Ponnurangam Kumaraguru. Credibility ranking of tweets during high impact events. In *Proceedings of the 1st Workshop on Privacy and Security in Online Social Media*, pages 2–8, Lyon, France, 2012.

[92] Maya R. Gupta and Yihua Chen. Theory and use of the EM algorithm. *Found. Trends Signal Process.*, 4(3):223–296, March 2011.

[93] Zoltán Gyöngyi, Hector Garcia-Molina, and Jan Pedersen. Combating Web spam with TrustRank. In *Proceedings of the Thirtieth International Conference on Very Large Data Bases*, VLDB '04, pages 576–587, Toronto, Canada, 2004.

[94] Chung-Wei Hang, Yonghong Wang, and Munindar P Singh. Operators for propagating trust and their evaluation in social networks. In *Proceedings of the 8th International Conference on Autonomous Agents and Multiagent Systems*, volume 2, pages 1025–1032, 2009.

[95] Trevor Hastie, Robert Tibshirani, and Jerome Friedman. *The Elements of Statistical Learning*. Springer Series in Statistics. New York, NY, 2001.

[96] Milena M Head and Khaled Hassanein. Trust in e-commerce: Evaluating the impact of third-party seals. *Quarterly Journal of Electronic Commerce*, 3:307–326, 2002.

[97] Jonathan L Herlocker, Joseph A Konstan, Loren G Terveen, and John T Riedl. Evaluating collaborative filtering recommender systems. *ACM Transactions on Information Systems (TOIS)*, 22(1):5–53, 2004.

[98] Seth Hettich and Michael J. Pazzani. Mining for proposal reviewers: Lessons learned at the national science foundation. In *Proceedings of the 12th ACM SIGKDD International Conference on Knowledge Discovery and Data Mining*, KDD '06, pages 862–871, Philadelphia, PA, 2006.

[99] Thomas Hofmann. Latent semantic models for collaborative filtering. *ACM Transactions on Information Systems (TOIS)*, 22(1):89–115, 2004.

[100] Carl I Hovland and Walter Weiss. The influence of source credibility on communication effectiveness. *Public opinion quarterly*, 15(4):635–650, 1951.

[101] Jeff Howe. The rise of crowdsourcing. *Wired Magazine*, 14(6):1–4, 2006.

[102] Panos Ipeirotis. Mechanical turk: Now with 40.92% spam. Behind Enemy Lines blog, 2010. http://www.behind-the-enemy-lines.com/2010/12/mechanical-turk-now-with-4092-spam.html.

[103] Athirai A. Irissappane, Siwei Jiang, and Jie Zhang. A biclustering-based approach to filter dishonest advisors in multi-criteria e-marketplaces. In *Proceedings of the 2014 International Conference on Autonomous Agents and Multi-Agent Systems*, AAMAS '14, pages 1385–1386, Paris, France, 2014.

[104] Mohsen Jamali and Martin Ester. Trustwalker: A random walk model for combining trust-based and item-based recommendation. In *Proceedings of the 15th ACM SIGKDD International Conference on Knowledge Discovery and Data Mining*, KDD '09, pages 397–406, Paris, France, 2009.

[105] Mohsen Jamali and Martin Ester. A matrix factorization technique with trust propagation for recommendation in social networks. In *Proceedings of the Fourth ACM Conference on Recommender Systems*, RecSys '10, pages 135–142, Barcelona, Spain, 2010.

[106] Susan Jamieson. Likert scales: How to (ab)use them. *Medical education*, 38(12):1217–1218, 2004.

[107] Dietmar Jannach, Markus Zanker, Alexander Felfernig, and Gerhard Friedrich. *Recommender systems: An introduction*. Cambridge University Press, Cambridge, United Kingdom, 2010.

[108] Carlos Jensen, John Davis, and Shelly Farnham. Finding others online: Reputation systems for social online spaces. In *Proceedings of the SIGCHI Conference on Human Factors in Computing Systems*, CHI '02, pages 447–454, Minneapolis, Minnesota, 2002.

[109] Nitin Jindal and Bing Liu. Opinion spam and analysis. In *Proceedings of the 2008 International Conference on Web Search and Data Mining*, WSDM '08, pages 219–230, Palo Alto, California, 2008.

[110] Thomas J Johnson and Barbara K Kaye. Using is believing: The influence of reliance on the credibility of online political information among politically interested Internet users. *Journalism & Mass Communication Quarterly*, 77(4):865–879, 2000.

[111] Thomas J Johnson and Barbara K Kaye. Webelievability: A path model examining how convenience and reliance predict online credibility. *Journalism & Mass Communication Quarterly*, 79(3):619–642, 2002.

[112] A Jøsang. Artificial reasoning with subjective logic. In *2nd Australian Workshop on Commonsense Reasoning*, Perth, Australia, 1997.

[113] A. Jøsang and J. Haller. Dirichlet reputation systems. In *Proceedings of the 2nd International Conference on Availability, Reliability and Security*, pages 112–119, Vienna, Austria, 2007.

[114] A. Jøsang, R. Hayward, and S. Pope. Trust network analysis with subjective logic. In *Proceedings of the 29th Australasian Computer Science Conference*, pages 85–94, Hobart, Australia, 2006.

[115] A. Jøsang and R. Ismail. The beta reputation system. In *Proceedings of the 15th Bled Electronic Commerce Conference*, pages 324–337, Bled, Slovenia, 2002.

[116] A. Josang, S. Marsh, and S. Pope. Exploring different types of trust propagation. *Lecture Notes in Computer Science*, 3986:179, 2006.

[117] Audun Jøsang, Roslan Ismail, and Colin Boyd. A survey of trust and reputation systems for online service provision. *Decision Support Systems*, 43(2):618–644, 2007.

[118] Michal Kakol, Michal Jankowski-Lorek, Katarzyna Abramczuk, Adam Wierzbicki, and Michele Catasta. On the subjectivity and bias of web content credibility evaluations. In *Proceedings of the 22nd International Conference on World Wide Web Companion*, WWW '13 Companion, pages 1131–1136, Rio de Janeiro, Brazil, 2013.

[119] Sepandar D. Kamvar, Mario T. Schlosser, and Hector Garcia-Molina. The eigentrust algorithm for reputation management in P2P networks. In *Proceedings of the 12th International Conference on World Wide Web*, WWW '03, pages 640–651, Budapest, Hungary, 2003.

[120] N. Karmarkar. A new polynomial-time algorithm for linear programming. *Combinatorica*, 4(4):373–395, 1984.

[121] J M Kleinberg. Authoritative sources in a hyperlinked environment. *Journal of the ACM*, 46(5):604–632, 1999.

[122] Joseph A Konstan. Introduction to recommender systems: Algorithms and evaluation. *ACM Transactions on Information Systems (TOIS)*, 22(1):1–4, 2004.

[123] Yehuda Koren. Collaborative filtering with temporal dynamics. *Communications of the ACM*, 53(4):89–97, 2010.

[124] Yehuda Koren, Robert Bell, and Chris Volinsky. Matrix factorization techniques for recommender systems. *Computer*, 42(8):30–37, 2009.

[125] Vassilis Kostakos. Is the crowd's wisdom biased? A quantitative analysis of three online communities. In *Computational Science and Engineering, 2009. CSE'09. International Conference on*, volume 4, pages 251–255, Vancouver, Canada, 2009. IEEE.

[126] Vijay Krishnan and Rashmi Raj. Web spam detection with anti-trust rank. In *The Second International Workshop on Adversarial Information Retrieval on the Web*, pages 37–40, Seattle, WA, 2006.

[127] Henry E Kyburg Jr. Bayesian and non-Bayesian evidential updating. *Artificial Intelligence*, 31(3):271–293, 1987.

[128] C. de Laat, G. Gross, L. Gommans, J. Vollbrecht, and D. Spence. Generic aaa architecture, rfc 2903. *Network Working Group*, 2000.

[129] Shyong K. Lam and John Riedl. Shilling recommender systems for fun and profit. In *Proceedings of the 13th International Conference on World Wide Web*, WWW '04, pages 393–402, New York, NY, 2004.

[130] Jaron   Lanier.   Digital   maoism: The   hazards of   the   new   online   collectivism.   *The*   *Edge*, http://www.edge.org/3rd_culture/lanier06/lanier06_index.html   (accessed July 16, 2014), 2006.

[131] Jaron Lanier. *You Are Not a Gadget*. Random House Digital, Inc., New York, NY, 2010.

[132] Harold Dwight Lasswell, Daniel Lerner, and Hans Speier. *Emergence of public opinion in the West*, volume 2. University Press of Hawaii, 1980.

[133] Hady W. Lauw, Ee-Peng Lim, and Ke Wang. Bias and controversy: Beyond the statistical deviation. In *Proceedings of the 12th ACM SIGKDD International Conference on Knowledge Discovery and Data Mining*, KDD '06, pages 625–630, Philadelphia, PA, 2006.

[134] Hady W. Lauw, Ee-Peng Lim, and Ke Wang. Summarizing review scores of "unequal" reviewers. In *Proceedings of the 2007 SIAM International Conference on Data Mining*, pages 539–544, Minneapolis, Minnesota, 2007.

[135] Hady W. Lauw, Ee-Peng Lim, and Ke Wang. Bias and controversy in evaluation systems. *TKDE*, 20(11):1490–1504, November 2008.

[136] Hady W. Lauw, Ee-Peng Lim, and Ke Wang. Quality and leniency in online collaborative rating systems. *TWEB*, 6(1):4:1–4:27, March 2012.

[137] HadyW. Lauw, Ee-Peng Lim, and Ke Wang. On mining rating dependencies in online collaborative rating networks. In Thanaruk Theeramunkong, Boonserm Kijsirikul, Nick Cercone, and Tu-Bao Ho, editors, *Advances in Knowledge Discovery and Data Mining*, volume 5476 of *Lecture Notes in Computer Science*, pages 1054–1061. Springer Berlin Heidelberg, 2009.

[138] Jonathan Lazar, Gabriele Meiselwitz, and Jinjuan Feng. *Understanding Web credibility: A synthesis of the research literature*, volume 1 of *Foundations and trends in human–computer interaction*. Now Publishers, Delft, The Netherlands, 2007.

[139] H Khac Le, J Pasternack, H Ahmadi, M Gupta, Y Sun, T Abdelzaher, J Han, D Roth, B Szymanski, and S Adali. Apollo: Towards factfinding in participatory sensing. In *Proceedings of the International Conference on Information Processing in Sensor Networks*, pages 129–130, Chicago, IL, 2011.

[140] Robert K Leik. A measure of ordinal consensus. *The Pacific Sociological Review*, 9(2):85–90, 1966.

[141] Daniel Lemire and Anna Maclachlan. Slope one predictors for online rating-based collaborative filtering. In *Proceedings of the Fifth SIAM International Conference on Data Mining*, volume 5, pages 1–5, Newport Beach, CA, 2005.

[142] R. Levien. Attack-resistant trust metrics. *Computing with Social Trust*, pages 121–132, 2008.

[143] J David Lewis and Andrew Weigert. Trust as a social reality. *Social Forces*, 63(4):967–985, 1985.

[144] Yanen Li, Jia Hu, ChengXiang Zhai, and Ye Chen. Improving one-class collaborative filtering by incorporating rich user information. In *Proceedings of the 19th ACM International Conference on Information and Knowledge Management*, CIKM '10, pages 959–968, Toronto, ON, 2010.

[145] Ee-Peng Lim, Viet-An Nguyen, Nitin Jindal, Bing Liu, and Hady Wirawan Lauw. Detecting product review spammers using rating behaviors. In *Proceedings of the 19th ACM International Conference on Information and Knowledge Management*, CIKM '10, pages 939–948, Toronto, ON, 2010.

[146] Sook Lim. College students' credibility judgments and heuristics concerning Wikipedia. *Information Processing & Management*, 49(2):405–419, 2013.

[147] Greg Linden, Brent Smith, and Jeremy York. Amazon. com recommendations: Item-to-item collaborative filtering. *Internet Computing, IEEE*, 7(1):76–80, 2003.

[148] Grzegorz Lissowski, Jacek Haman, and Mikołaj Jasiński. *Podstawy statystyki dla socjologów*. Wydawnictwo Naukowe Scholar, 2008.

[149] S. Liu, J. Zhang, C. Miao, Y.L. Theng, and A.C. Kot. iCLUB: An integrated clustering-based approach to improve the robustness of reputation systems. In *Proceedings of the International Conference on Autonomous Agents and Multiagent Systems*, pages 1151–1152, Taipei, Taiwan, 2011.

[150] X. Liu, A. Datta, K. Rzadca, and E.P. Lim. Stereotrust: A group based personalized trust model. In *Proceedings of the 18th ACM conference on Information and Knowledge Management*, pages 7–16, Hong Kong, China, 2009.

[151] X. Liu, T. Kaszuba, R. Nielek, A. Datta, and A. Wierzbicki. Using stereotypes to identify risky transactions in Internet auctions. In *Proceedings of the IEEE Second International Conference on Social Computing*, pages 513–520, Minneapolis, MN, 2010.

[152] X. Liu, G. Tredan, and A. Datta. Metatrust: Discriminant analysis of local information for global trust assessment. In *Proceedings of the 10th International Conference on Autonomous Agents and Multiagent Systems*, pages 1071–1072, Taipei, Taiwan, 2011.

[153] X. Liu, G. Tredan, and A. Datta. A generic trust framework for large-scale open systems using machine learning. *Computational Intelligence*, published online: October 29, 2013.

[154] Xin Liu and Anwitaman Datta. Contextual trust aided enhancement of data availability in peer-to-peer backup storage systems. *Journal of Network and Systems Management*, 20:200–225, 2012.

[155] Xin Liu and Anwitaman Datta. Modeling context aware dynamic trust using hidden Markov model. In *Proceedings of the Twenty-Sixth Conference on Artificial Intelligence*, pages 1938–1944, Toronto, Canada, 2012.

[156] Teun Lucassen and Jan Maarten Schraagen. The influence of source cues and topic familiarity on credibility evaluation. *Computers in Human Behavior*, 29(4):1387–1392, 2013.

[157] Hao Ma, Irwin King, and Michael R Lyu. Learning to recommend with social trust ensemble. In *Proceedings of the 32nd international ACM SIGIR conference on Research and development in information retrieval*, pages 203–210. ACM, 2009.

[158] Hao Ma, Michael R. Lyu, and Irwin King. Learning to recommend with trust and distrust relationships. In *Proceedings of the Third ACM Conference on Recommender Systems*, RecSys '09, pages 189–196, New York, NY, 2009.

[159] Hao Ma, Haixuan Yang, Michael R. Lyu, and Irwin King. SoRec: Social recommendation using probabilistic matrix factorization. In *Proceedings of the 17th ACM Conference on Information and Knowledge Management*, CIKM '08, pages 931–940, Napa Valley, California, 2008.

[160] Hao Ma, Dengyong Zhou, Chao Liu, Michael R. Lyu, and Irwin King. Recommender systems with social regularization. In *Proceedings of the*

Fourth ACM International Conference on Web Search and Data Mining, WSDM '11, pages 287–296, Hong Kong, China, 2011.

[161] Zaki Malik, Ihsan Akbar, and Athman Bouguettaya. Web services reputation assessment using a hidden Markov model. In *Proceedings of the 7th International Joint Conference on Service-Oriented Computing*, ICSOC-ServiceWave '09, pages 576–591, Stockholm, Sweden, 2009.

[162] Zaki Malik and Athman Bouguettaya. Rateweb: Reputation assessment for trust establishment among Web services. *The VLDB Journal—The International Journal on Very Large Data Bases*, 18(4):885–911, 2009.

[163] Daniel W. Manchala. Trust metrics, models and protocols for electronic commerce transactions. In *Proceedings of the The 18th International Conference on Distributed Computing Systems*, ICDCS '98, pages 312–321, Amsterdam, The Netherlands, 1998. IEEE Computer Society.

[164] Stephen P. Marsh. *Formalising Trust as a Computational Concept*. PhD thesis, University of Stirling, 1994.

[165] Sergio Marti and Hector Garcia-Molina. Limited reputation sharing in P2P systems. In *Proceedings of the 5th ACM Conference on Electronic Commerce*, pages 91–101, New York, NY, 2004.

[166] P. Massa and P. Avesani. Trust-aware recommender systems. In *Proceedings of the 2007 ACM conference on Recommender systems*, pages 17–24, Minneapolis, MN, 2007.

[167] Paolo Massa and Paolo Avesani. Trust-aware collaborative filtering for recommender systems. In Robert Meersman and Zahir Tari, editors, *On the Move to Meaningful Internet Systems 2004: CoopIS, DOA, and ODBASE*, pages 492–508. Springer, Berlin Heidelberg, 2004.

[168] Paolo Massa and Paolo Avesani. Controversial users demand local trust metrics: An experimental study on epinions.com community. In *Proceedings of the 20th National Conference on Artificial Intelligence - Volume 1*, AAAI'05, pages 121–126, Pittsburgh, Pennsylvania, 2005.

[169] Paolo Massa and Paolo Avesani. Trust metrics on controversial users: Balancing between tyranny of the majority. *International Journal on Semantic Web and Information Systems (IJSWIS)*, 3(1):39–64, 2007.

[170] Paolo Massa and Bobby Bhattacharjee. Using trust in recommender systems: An experimental analysis. In Christian Jensen, Stefan Poslad, and Theo Dimitrakos, editors, *Trust Management*, pages 221–235. Springer, Berlin Heidelberg, 2004.

[171] Mark T. Maybury and Wolfgang Wahlster, editors. *Readings in Intelligent User Interfaces*. Morgan Kaufmann Publishers Inc., San Francisco, CA, USA, 1998.

[172] Roger C Mayer, James H Davis, and F David Schoorman. An integrative model of organizational trust. *Academy of Management Review*, 20(3):709–734, 1995.

[173] Andrew McCallum and Kamal Nigam. A comparison of event models for naive Bayes text classification. In *AAAI-98 Workshop on Learning for Text Categorization*, volume 752, pages 41–48, Madison, Wisconsin, 1998.

[174] D.H. McKnight and N.L. Chervany. The meanings of trust. Technical report, University of Minnesota, Management Information Systems Research Center, 1996.

[175] Loubna Mekouar, Youssef Iraqi, and Raouf Boutaba. Reputation-based trust management in Peer-to-Peer systems: Axonomy and anatomy. In Xuemin Shen, Heather Yu, John Buford, and Mursalin Akon, editors, *Handbook of Peer-to-Peer Networking*, pages 689–732. Springer, New York, NY, 2010.

[176] A. Menon and C. Elkan. Link prediction via matrix factorization. *Machine Learning and Knowledge Discovery in Databases*, pages 437–452, 2011.

[177] Miriam Metzger. Understanding how Internet users make sense of credibility: A review of the state of our knowledge and recommendations for theory, policy, and practice. In *Symposium on Internet Credibility and the User*, pages 1–28, Seattle, Washington, 2005.

[178] Miriam J Metzger. Making sense of credibility on the Web: Models for evaluating online information and recommendations for future research. *Journal of the American Society for Information Science and Technology*, 58(13):2078–2091, 2007.

[179] Miriam J Metzger and Andrew J Flanagin. Credibility and trust of information in online environments: The use of cognitive heuristics. *Journal of Pragmatics*, 59:210–220, 2013.

[180] Miriam J Metzger, Andrew J Flanagin, and Ryan B Medders. Social and heuristic approaches to credibility evaluation online. *Journal of Communication*, 60(3):413–439, 2010.

[181] Miriam J Metzger, Andrew J Flanagin, and Lara Zwarun. College student Web use, perceptions of information credibility, and verification behavior. *Computers & Education*, 41(3):271–290, 2003.

[182] Abhinav Mishra and Rajeev Rastogi. Semi-supervised correction of biased comment ratings. In *Proceedings of the 21st International Conference on World Wide Web*, WWW '12, pages 181–190, Lyon, France, 2012.

[183] Bamshad Mobasher, Robin Burke, and J. J. Sandvig. Model-based collaborative filtering as a defense against profile injection attacks. In *Proceedings of the 21st National Conference on Artificial Intelligence - Volume 2*, AAAI'06, pages 1388–1393, Boston, Massachusetts, 2006.

[184] Marie E. G. Moe, Mozhgan Tavakolifard, and Svein J. Knapskog. Learning trust in dynamic multiagent environments using HMMs. In *Proceedings of The 13th Nordic Workshop on Secure IT Systems*, Copenhagen, Denmark, 2008.

[185] Marie E.G. Moe, Bjarne E. Helvik, and Svein J. Knapskog. Tsr: Trust-based secure manet routing using HMMs. In *Proceedings of the 4th ACM Symposium on QoS and Security for Wireless and Mobile Networks*, pages 83–90, Vancouver, Canada, 2008.

[186] MarieE.G. Moe, BjarneE. Helvik, and SveinJ. Knapskog. Comparison of the beta and the hidden Markov models of trust in dynamic environments. In Elena Ferrari, Ninghui Li, Elisa Bertino, and Yuecel Karabulut, editors, *Trust Management III*, volume 300 of *IFIP Advances in Information and Communication Technology*, pages 283–297. Springer, Berlin Heidelberg, 2009.

[187] Raymond J. Mooney and Loriene Roy. Content-based book recommending using learning for text categorization. In *Proceedings of the Fifth ACM Conference on Digital Libraries*, DL '00, pages 195–204, San Antonio, TX, 2000.

[188] Alex Morales, Huan Sun, and Xifeng Yan. Synthetic review spamming and defense. In *Proceedings of the 22Nd International Conference on World Wide Web Companion*, WWW '13 Companion, pages 155–156, Rio de Janeiro, Brazil, 2013.

[189] Meredith Ringel Morris, Scott Counts, Asta Roseway, Aaron Hoff, and Julia Schwarz. Tweeting is believing?: Understanding microblog credibility perceptions. In *Proceedings of the ACM Conference on Computer Supported Cooperative Work*, CSCW '12, pages 441–450, Seattle, Washington, 2012. ACM.

[190] Arjun Mukherjee, Abhinav Kumar, Bing Liu, Junhui Wang, Meichun Hsu, Malu Castellanos, and Riddhiman Ghosh. Spotting opinion spammers using behavioral footprints. In *Proceedings of the 19th ACM SIGKDD International Conference on Knowledge Discovery and Data Mining*, KDD '13, pages 632–640, Chicago, IL, 2013.

[191] Arjun Mukherjee, Bing Liu, Junhui Wang, Natalie Glance, and Nitin Jindal. Detecting group review spam. In *Proceedings of the 20th International Conference Companion on World Wide Web*, WWW '11, pages 93–94, Hyderabad, India, 2011.

[192] Arjun Mukherjee, Vivek Venkataraman, Bing Liu, and Natalie Glance. What Yelp fake review filter might be doing. In *Proceedings of the 7th International AAAI Conference on Weblogs and Social Media*, pages 409–418, Ann Arbor, Michigan, 2013.

[193] Kevin P. Murphy. *Machine Learning: A Probabilistic Perspective*. The MIT Press, Cambridge, Massachusetts, 2012.

[194] Clark Naeemah. Trust me! Wikipedias credibility among college students. *International Journal of Instructional Media*, 38(1):27–36, 2011.

[195] Radoslaw Nielek, Aleksander Wawer, Michal Jankowski-Lorek, and Adam Wierzbicki. Temporal, cultural and thematic aspects of Web credibility. In *Proceedings of the 5th International Conference on Social Informatics*, pages 419–428, Kyoto, Japan, 2013.

[196] V Novak, I Perfilieva, and J Mockof. *Mathematical Principles of Fuzzy Logic*. Kluwer Academic Publishers, Berlin, Germany, 1999.

[197] John O'Donovan and Barry Smyth. Trust in recommender systems. In *Proceedings of the 10th International Conference on Intelligent User Interfaces*, pages 167–174. ACM, 2005.

[198] Alexandra Olteanu, Stanislav Peshterliev, Xin Liu, and Karl Aberer. Web credibility: Features exploration and credibility prediction. In *Proceedings of the 35th European Conference on Advances in Information Retrieval*, ECIR'13, pages 557–568, Moscow, Russia, 2013.

[199] Myle Ott, Yejin Choi, Claire Cardie, and Jeffrey T Hancock. Finding deceptive opinion spam by any stretch of the imagination. In *ACL HLT*, pages 309–319. Association for Computational Linguistics, 2011.

[200] L. Page, S. Brin, R. Motwani, and T. Winograd. The pagerank citation ranking: Bringing order to the web. Technical Report 1999-66, Stanford InfoLab, November 1999.

[201] Rong Pan, Yunhong Zhou, Bin Cao, Nathan N Liu, Rajan Lukose, Martin Scholz, and Qiang Yang. One-class collaborative filtering. In *Proceedings of the IEEE International Conference on Data Mining*, pages 502–511, 2008.

[202] Jeff Pasternack and Dan Roth. Knowing what to believe (when you already know something). In *Proceedings of the 23rd International Conference on Computational Linguistics*, COLING '10, pages 877–885, Beijing, China, 2010.

[203] Jeff Pasternack and Dan Roth. Generalized fact-finding. In *Proceedings of the 20th International Conference Companion on World Wide Web*, WWW '11, pages 99–100, Hyderabad, India, 2011.

[204] Jeff Pasternack and Dan Roth. Making better informed trust decisions with generalized fact-finding. In *Proceedings of the Twenty-Second International Joint Conference on Artificial Intelligence - Volume Volume Three*, IJCAI'11, pages 2324–2329, Barcelona, Spain, 2011.

[205] Jeff Pasternack and Dan Roth. Latent credibility analysis. In *Proceedings of the 22Nd International Conference on World Wide Web*, WWW '13, pages 1009–1020, Rio de Janeiro, Brazil, 2013.

[206] MichaelJ. Pazzani and Daniel Billsus. Content-based recommendation systems. In Peter Brusilovsky, Alfred Kobsa, and Wolfgang Nejdl, editors, *The Adaptive Web*, volume 4321 of *Lecture Notes in Computer Science*, pages 325–341. Springer, Berlin Heidelberg, 2007.

[207] David M. Pennock, Eric Horvitz, Steve Lawrence, and C. Lee Giles. Collaborative filtering by personality diagnosis: A hybrid memory and model-based approach. In *Proceedings of the 16th Conference on Uncertainty in Artificial Intelligence*, UAI '00, pages 473–480, Stanford, CA, 2000.

[208] Richard E Petty and John T Cacioppo. *Attitudes and Persuasion: Classic and Contemporary Approaches*. Westview Press, Boulder, CO, 1996.

[209] RichardE. Petty and JohnT. Cacioppo. The elaboration likelihood model of persuasion. In *Communication and Persuasion*, Springer Series in Social Psychology, pages 1–24. Springer, New York, NY, 1986.

[210] Selwyn Piramuthu, Gaurav Kapoor, Wei Zhou, and Sjouke Mauw. Input online review data and related bias in recommender systems. *Decision Support Systems*, 53(3):418–424, 2012.

[211] Hoifung Poon and Pedro Domingos. Joint inference in information extraction. In *Proceedings of the 22Nd National Conference on Artificial Intelligence - Volume 1*, AAAI'07, pages 913–918, Vancouver, Canada, 2007.

[212] V. Punyakanok, D. Roth, W. Yih, and D. Zimak. Learning and inference over constrained output. In *International Joint Conference on Artificial Intelligence*, volume 19, 2005.

[213] Christopher Puto. Adding asymmetrically dominated alternatives: Violations of regularity and the similarity hypothesis. *Journal of Consumer Research*, 9(1):90–98, 1982.

[214] J. R. Quinlan. Learning with continuous classes. In *Proceedings of the Australian Joint Conference on Artificial Intelligence*, pages 343–348, Hobart, Tasmania, 1992.

[215] Sindhu Raghavan, Suriya Gunasekar, and Joydeep Ghosh. Review quality aware collaborative filtering. In *Proceedings of the Sixth ACM Conference on Recommender Systems*, RecSys '12, pages 123–130, Dublin, Ireland, 2012.

[216] Vikas C Raykar and Shipeng Yu. Eliminating spammers and ranking annotators for crowdsourced labeling tasks. *The Journal of Machine Learning Research*, 13:491–518, 2012.

[217] Steven Reece, Alex Rogers, Stephen Roberts, and Nicholas R. Jennings. Rumours and reputation: Evaluating multi-dimensional trust within a decentralised reputation system. In *Proceedings of the 6th International Joint Conference on Autonomous Agents and Multiagent Systems*, AAMAS '07, pages 165:1–165:8, Honolulu, Hawaii, 2007.

[218] K. Regan, P. Poupart, and R. Cohen. Bayesian reputation modeling in e-marketplaces sensitive to subjectivity, deception and change. In *Proceedings of the National Conference on Artificial Intelligence*, 2006.

[219] Paul Resnick, Ko Kuwabara, Richard Zeckhauser, and Eric Friedman. Reputation systems. *Communications of the ACM*, 43(12):45–48, 2000.

[220] Paul Resnick and Hal R Varian. Recommender systems. *Communications of the ACM*, 40(3):56–58, 1997.

[221] Paul Resnick, Richard Zeckhauser, John Swanson, and Kate Lockwood. The value of reputation on eBay: A controlled experiment. *Experimental Economics*, 9(2):79–101, 2006.

[222] Soo Young Rieh and Brian Hilligoss. College students' credibility judgments in the information-seeking process. *Digital Media, Youth, and Credibility*, pages 49–72, 2008.

[223] Tracy Riggs and Robert Wilensky. An algorithm for automated rating of reviewers. In *Proceedings of the 1st ACM/IEEE-CS Joint Conference on Digital Libraries*, JCDL '01, pages 381–387, Roanoke, Virginia, 2001.

[224] Tony Rimmer and David Weaver. Different questions, different answers? Media use and media credibility. *Journalism & Mass Communication Quarterly*, 64(1):28–44, 1987.

[225] Magnus Roos, Jörg Rothe, Joachim Rudolph, Björn Scheuermann, and Dietrich Stoyan. A statistical approach to calibrating the scores of biased reviewers: The linear vs. the nonlinear model. In *Proceedings of the 6th Multidisciplinary Workshop on Advances in Preference Handling*, pages 30–35, Montpellier, France, 2012.

[226] Magnus Roos, Jörg Rothe, and Björn Scheuermann. How to calibrate the scores of biased reviewers by quadratic programming. In *Proceedings*

of the Twenty-Fifth Conference on Artificial Intelligence, pages 255–260, San Francisco, CA, 2011.

[227] D. Roth and W. Yih. Global inference for entity and relation identification via a linear programming formulation. *Introduction to Statistical Relational Learning*, pages 553–580, 2007.

[228] Dan Roth and Wentau Yih. A linear programming formulation for global inference in natural language tasks. In *HLT-NAACL 2004 Workshop: Eighth Conference on Computational Natural Language Learning*, pages 1–8, Boston, MA, 2004.

[229] Stuart Russell and Peter Norvig. *Artificial Intelligence: A Modern Approach*. Prentice Hall, Upper Saddle River, NJ, second edition, 2003.

[230] Paul Ruvolo, Jacob Whitehill, and Javier R Movellan. Exploiting structure in crowdsourcing tasks via latent factor models. Technical report, Machine Perception Laboratory, 2010.

[231] Jordi Sabater and Carles Sierra. Review on computational trust and reputation models. *Artificial Intelligence Review*, 24(1):33–60, September 2005.

[232] Badrul Sarwar, George Karypis, Joseph Konstan, and John Riedl. Application of dimensionality reduction in recommender system—A case study. Technical report, University of Minnesota, 2000.

[233] Badrul Sarwar, George Karypis, Joseph Konstan, and John Riedl. Item-based collaborative filtering recommendation algorithms. In *Proceedings of the 10th international conference on World Wide Web*, pages 285–295, Hong Kong, China, 2001.

[234] J. Ben Schafer, Joseph Konstan, and John Riedl. Recommender systems in e-commerce. In *Proceedings of the 1st ACM Conference on Electronic Commerce*, EC '99, pages 158–166, Denver, Colorado, 1999.

[235] Andrew I. Schein, Alexandrin Popescul, Lyle H. Ungar, and David M. Pennock. Methods and metrics for cold-start recommendations. In *Proceedings of the 25th Annual International ACM SIGIR Conference on Research and Development in Information Retrieval*, SIGIR '02, pages 253–260, Tampere, Finland, 2002.

[236] Michael Schillo, Petra Funk, and Michael Rovatsos. Using trust for detecting deceitful agents in artificial societies. *Applied Artificial Intelligence*, 14(8):825–848, 2000.

[237] Julia Schwarz and Meredith Morris. Augmenting web pages and search results to support credibility assessment. In *Proceedings of the SIGCHI Conference on Human Factors in Computing Systems*, pages 1245–1254, Vancouver, Canada, 2011. ACM.

[238] Aaditeshwar Seth, Jie Zhang, and Robin Cohen. A subjective credibility model for participatory media. In *Workshop on Recommender Systems, the Twenty-Third AAAI Conference on Artificial Intelligence*, pages 66–77, Chicago, IL, 2008.

[239] G Shafer. *A mathematical Theory of Evidence*. Princeton University Press, Princeton, NJ, 1976.

[240] Bracha Shapira. *Recommender Systems Handbook*. Springer, New York, NY, 2011.

[241] Xiao-Liang Shen, Christy M. K. Cheung, and Matthew K. O. Lee. What leads students to adopt information from Wikipedia? An empirical investigation into the role of trust and information usefulness. *British Journal of Educational Technology*, 44:502–517, 2012.

[242] Victor S Sheng, Foster Provost, and Panagiotis G Ipeirotis. Get another label? Improving data quality and data mining using multiple, noisy labelers. In *Proceedings of the 14th ACM SIGKDD International Conference on Knowledge Discovery and Data Mining*, pages 614–622, Las Vegas, Nevada, USA, 2008. ACM.

[243] Wanita Sherchan, Surya Nepal, and Cecile Paris. A survey of trust in social networks. *ACM Computing Surveys (CSUR)*, 45(4):47, 2013.

[244] Riyaz Sikora and Liangjun You. Effect of reputation mechanisms and ratings biases on traders' behavior in online marketplaces. *Journal of Organizational Computing and Electronic Commerce*, 24:58–73, 2014.

[245] Riyaz T Sikora and Kriti Chauhan. Estimating sequential bias in online reviews: A Kalman filtering approach. *Knowledge-Based Systems*, 27:314–321, 2012.

[246] Rashmi R Sinha and Kirsten Swearingen. Comparing recommendations made by online systems and friends. In *DELOS workshop: personalisation and recommender systems in digital libraries*, volume 106, 2001.

[247] Parikshit Sondhi, V. G. Vinod Vydiswaran, and Cheng Xiang Zhai. Reliability prediction of Webpages in the medical domain. In *Proceedings of the 34th European Conference on Advances in Information Retrieval*, ECIR'12, pages 219–231, Barcelona, Spain, 2012.

[248] Shanshan Song, Kai Hwang, Runfang Zhou, and Yu-Kwong Kwok. Trusted P2P transactions with fuzzy reputation aggregation. *Internet Computing, IEEE*, 9(6):24–34, 2005.

[249] Don W Stacks and Michael Brian Salwen. *An integrated approach to communication theory and research*. Routledge, New York, NY, 2009.

[250] Jessica Staddon and Richard Chow. Detecting reviewer bias through Web-based association mining. In *Proceedings of the 2Nd ACM Workshop on Information Credibility on the Web*, WICOW '08, pages 5–10, Napa Valley, CA, 2008.

[251] Keith Stamm and Ric Dube. The relationship of attitudinal components to trust in media. *Communication Research*, 21(1):105–123, 1994.

[252] X. Su and Taghi M. Khoshgoftaar. A survey of collaborative filtering techniques. *Advances in Artificial Intelligence*, 2009:1–20, January 2009.

[253] Fabian M. Suchanek, Mauro Sozio, and Gerhard Weikum. Sofie: A self-organizing framework for information extraction. In *Proceedings of the 18th International Conference on World Wide Web*, WWW '09, pages 631–640, Madrid, Spain, 2009.

[254] James Surowiecki. *The Wisdom of Crowds*. Random House Digital, Inc., New York, NY, 2005.

[255] S. Tadelis. Firm reputation with hidden information. *Economic Theory*, 21:635–651, 2003.

[256] Nassim Nicholas Taleb. The fourth quadrant: A map of the limits of statistics. *An Edge Original Essay*, available online: http://edge.org/conversation/the-fourth-quadrant-a-map-of-the-limits-of-statistics, 2008.

[257] Jiliang Tang, Huiji Gao, Xia Hu, and Huan Liu. Exploiting homophily effect for trust prediction. In *Proceedings of the sixth ACM international conference on Web Search and Data Mining*, pages 53–62, Rome, Italy, 2013. ACM.

[258] Jiliang Tang, Huiji Gao, and Huan Liu. mTrust: Discerning multi-faceted trust in a connected world. In *Proceedings of the fifth ACM international conference on Web Search and Data Mining*, pages 93–102, Seattle, WA, 2012.

[259] Jiliang Tang, Huiji Gao, Huan Liu, and Atish Das Sarma. etrust: Understanding trust evolution in an online world. In *Proceedings of the 18th ACM SIGKDD international conference on Knowledge Discovery and Data Mining*, pages 253–261, Beijing, China, 2012.

[260] Jiliang Tang, Xia Hu, Huiji Gao, and Huan Liu. Exploiting local and global social context for recommendation. In *Proceedings of the Twenty-Third International Joint Conference on Artificial Intelligence*, IJCAI'13, pages 2712–2718, Beijing, China, 2013.

[261] Jiliang Tang, Xia Hu, and Huan Liu. Social recommendation: A review. *Social Network Analysis and Mining*, 3(4):1113–1133, 2013.

[262] Jiliang Tang and Huan Liu. Coselect: Feature selection with instance selection for social media data. In *Proceedings of the 13th SIAM International Conference on Data Mining*, pages 695–703, 2013.

[263] William J Tastle and Mark J Wierman. Consensus and dissention: A measure of ordinal dispersion. *International Journal of Approximate Reasoning*, 45(3):531–545, 2007.

[264] W.T.L. Teacy, J. Patel, N.R. Jennings, and M. Luck. TRAVOS: Trust and reputation in the context of inaccurate information sources. *Autonomous Agents and Multi-Agent Systems*, 12(2):183–198, 2006.

[265] Loren Terveen and Will Hill. Beyond recommender systems: Helping people help each other. *HCI in the New Millennium*, 1:487–509, 2001.

[266] Shari Trewin. Knowledge-based recommender systems. *Encyclopedia of Library and Information Science*, 69(Supplement 32):69, 2000.

[267] S Tseng and B J Fogg. Credibility and computing technology. *Communications of the ACM*, 42(5):39–44, 1999.

[268] Amos Tversky and Daniel Kahneman. Belief in the law of small numbers. *Psychological Bulletin*, 76(2):105, 1971.

[269] Amos Tversky and Daniel Kahneman. Judgment under uncertainty: Heuristics and biases. *Science*, 185(4157):1124–1131, 1974.

[270] Cees Van der Eijk. Measuring agreement in ordered rating scales. *Quality and Quantity*, 35(3):325–341, 2001.

[271] Patricia Victor, Chris Cornelis, Martine De Cock, and Ankur Teredesai. Trust- and distrust-based recommendations for controversial reviews. *Intelligent Systems, IEEE*, 26(1):48–55, Jan 2011.

[272] Patricia Victor, Chris Cornelis, Martine De Cock, and Ankur M Teredesai. A comparative analysis of trust-enhanced recommenders for controversial items. In *Proceedings of the International AAAI Conference on Weblogs and Social Media*, pages 342–345, San Jose, CA, 2009.

[273] Patricia Victor, Martine De Cock, and Chris Cornelis. Trust and recommendations. In *Recommender Systems Handbook*, pages 645–675. Springer, New York, NY, 2011.

[274] Luis von Ahn and Laura Dabbish. Labeling images with a computer game. In *Proceedings of the SIGCHI Conference on Human Factors in Computing Systems*, CHI '04, pages 319–326, Vienna, Austria, 2004.

[275] Luis Von Ahn, Benjamin Maurer, Colin McMillen, David Abraham, and Manuel Blum. reCAPTCHA: Human-based character recognition via Web security measures. *Science*, 321(5895):1465–1468, 2008.

[276] Le-Hung Vu, Manfred Hauswirth, and Karl Aberer. Qos-based service selection and ranking with trust and reputation management. In *Proceedings of the 2005 Confederated International Conference on On the Move to Meaningful Internet Systems*, OTM'05, pages 466–483, Agia Napa, Cyprus, 2005.

[277] Frank Edward Walter, Stefano Battiston, and Frank Schweitzer. A model of a trust-based recommendation system on a social network. *Autonomous Agents and Multi-Agent Systems*, 16(1):57–74, 2008.

[278] Jun Wang, Arjen P De Vries, and Marcel JT Reinders. Unifying user-based and item-based collaborative filtering approaches by similarity fusion. In *Proceedings of the 29th Annual International ACM SIGIR Conference on Research and Development in Information Retrieval*, pages 501–508, Seattle, WA, 2006.

[279] Yan Wang, Kwei-Jay Lin, Duncan S Wong, and Vijay Varadharajan. Trust management towards service-oriented applications. *Service Oriented Computing and Applications*, 3(2):129–146, 2009.

[280] Yao Wang and Julita Vassileva. Trust and reputation model in peer-to-peer networks. In *Proceedings of the Third International Conference on Peer-to-Peer Computing*, pages 150–157, Linköping, Sweden, 2003.

[281] Yao Wang, Jie Zhang, and Julita Vassileva. A super-agent based framework for reputation management and community formation in decentralized systems. *Computational Intelligence*, published online, DOI: 10.1111/coin.12026, 2014.

[282] Wayne Wanta and Yu-Wei Hu. The effects of credibility, reliance, and exposure on media agenda-setting: A path analysis model. *Journalism & Mass Communication Quarterly*, 71(1):90–98, 1994.

[283] Wei Wei, Kam Tong Chan, Irwin King, and Jimmy Ho-Man Lee. Rate: A review of reviewers in a manuscript review process. In *Proceedings of IEEE/WIC/ACM International Conference on Web Intelligence and Intelligent Agent Technology*, pages 204–207, Sydney, Australia, 2008.

[284] J. Weise. Public key infrastructure overview. *Sun Blueprints Online*, available at http://www.sun.com/blueprints/0801/publickey.pdf, 2001.

[285] J. Weng, C.Y. Miao, and Angela Goh. An entropy-based approach to protecting rating systems from unfair testimonies. *IEICE—Transactions on Information and Systems*, E89-D:2502–2511, 2006.

[286] A. Whitby, A. Jøsang, and J. Indulska. Filtering out unfair ratings in Bayesian reputation systems. In *Proceedings of the International Joint Conference Autonomous Agents and Multiagent Systems Workshop on Trust in Agent Societies*, pages 106–117, New York, NY, 2004.

[287] Fei Wu and Daniel S. Weld. Autonomously semantifying wikipedia. In *Proceedings of the Sixteenth ACM Conference on Conference on Information and Knowledge Management*, CIKM '07, pages 41–50, Lisbon, Portugal, 2007.

[288] L. Xiong and L. Liu. Peertrust: Supporting reputation-based trust for peer-to-peer electronic communities. *IEEE Transactions on Knowledge and Data Engineering*, 16(7):843–857, July 2004.

[289] Yuan Yao, Hanghang Tong, Xifeng Yan, Feng Xu, and Jian Lu. Matri: A multi-aspect and transitive trust inference model. In *Proceedings of the 22nd International Conference on World Wide Web*, pages 1467–1476, Rio de Janeiro, Brazil, 2013.

[290] Xiaoxin Yin, Jiawei Han, and Philip S. Yu. Truth discovery with multiple conflicting information providers on the web. *IEEE Transactions on Knowledge and Data Engineering*, 20(6):796–808, 2008.

[291] Bin Yu and Munindar P. Singh. Detecting deception in reputation management. In *Proceedings of the Second International Joint Conference on Autonomous Agents and Multiagent Systems*, AAMAS '03, pages 73–80, Melbourne, Australia, 2003.

[292] Haifeng Yu, Michael Kaminsky, Phillip B Gibbons, and Abraham Flaxman. Sybilguard: Defending against sybil attacks via social networks. *ACM SIGCOMM Computer Communication Review*, 36(4):267–278, 2006.

[293] G. Zacharia. *Collaborative Reputation Mechanisms for Online Communities*. PhD thesis, Massachusetts Institute of Technology, Boston, MA, 1999.

[294] Giorgos Zacharia and Pattie Maes. Trust management through reputation mechanisms. *Applied Artificial Intelligence*, 14(9):881–907, 2000.

[295] Reza Zafarani, Mohammad Ali Abbasi, and Huan Liu. *Social Media Mining: An Introduction*. Cambridge University Press, Cambridge, United Kingdom, 2014.

[296] Honglei Zeng, Maher A. Alhossaini, Li Ding, Richard Fikes, and Deborah L. McGuinness. Computing trust from revision history. In *Proceedings of the 2006 International Conference on Privacy, Security and Trust: Bridge the Gap Between PST Technologies and Business Services*, PST '06, pages 1–8, Markham, Canada, 2006.

[297] Huanhuan Zhang, Chang Xu, and Jie Zhang. Exploiting trust and distrust information to combat sybil attack in online social networks. In *Proceedings of the 8th IFIP WG 11.11 International Conference on Trust Management*, pages 77–92, Singapore, 2014.

[298] J. Zhang and R. Cohen. Evaluating the trustworthiness of advice about seller agents in e-marketplaces: A personalized approach. *Electronic Commerce Research and Applications*, 7(3):330–340, 2008.

[299] Runfang Zhou and Kai Hwang. Powertrust: A robust and scalable reputation system for trusted peer-to-peer computing. *IEEE Transactions on Parallel and Distributed Systems*, 18(4):460–473, 2007.

[300] Runfang Zhou, Kai Hwang, and Min Cai. GossipTrust for fast reputation aggregation in peer-to-peer networks. *IEEE Transactions on Knowledge and Data Engineering*, 20(9):1282–1295, September 2008.

[301] Cai-Nicolas Ziegler. *Towards Decentralized Recommender Systems*. PhD thesis, University of Freiburg, Freiburg, Switzerland, 2005.

[302] Cai-Nicolas Ziegler and J. Golbeck. Investigating correlations of trust and interest similarity - do birds of a feather really flock together? *Decision Support Systems*, 42(3):1111–1136, 2005.

[303] Cai-Nicolas Ziegler and Jennifer Golbeck. Investigating interactions of trust and interest similarity. *Decision Support System*, 43(2):460–475, 2007.

[304] Cai-Nicolas Ziegler and Georg Lausen. Propagation models for trust and distrust in social networks. *Information Systems Frontiers*, 7(4–5):337–358, 2005.

Index